# Lessons from Hurricane Ike

# Lessons *from* Hurricane Ike

Edited by PHILIP B. BEDIENT

Texas A&M University Press • *College Station*

Library of Congress
Cataloging-in-Publication Data

   Lessons from Hurricane Ike / edited by
Philip B. Bedient.—1st ed.
     p. cm.
  Includes index.
  ISBN-13: 978-1-60344-588-7
(flexbound (with flaps) : alk. paper)
  ISBN-10: 1-60344-588-9
(flexbound (with flaps) : alk. paper)
  ISBN-13: 978-1-60344-736-2 (e-book)
  ISBN-10: 1-60344-736-9 (e-book)
  1. Hurricane Ike, 2008.   2. Hurricanes—
Texas—Gulf Coast.   3. Hurricane protection—
Texas—Gulf Coast.   4. Emergency management—
Texas—Gulf Coast.   5. Storm surges—Texas—
Gulf Coast.   I. Bedient, Philip B., 1948–
  HV636 2008 .T4 L47   2012
  363.34'922097641—dc23
  2011049844

# Contents

# Preface

As a young boy growing up on the west coast of Florida, I was always impressed by the sheer power of Mother Nature, especially the intense severe storms and major hurricanes that would come along every summer to interrupt our lives. I went on to obtain degrees in physics and environmental engineering at the University of Florida and developed my skills as a hydrologic engineer with a particular interest in flood prediction and warning. In 2001, Tropical Storm Allison was the most devastating urban flood in US history. My work with the Texas Medical Center and others began the research that I now conduct on severe storms in the Gulf Coast.

In 2005, shortly after hurricanes Katrina and Rita, I organized a technical meeting at Rice University to discuss the aftermath on the Gulf Coast. That evening over dinner I looked around a table of approximately fifteen researchers from Rice, the University of Texas, Texas A&M, the University of Houston, and Louisiana State University and realized that I had gathered the premier researchers equipped with the knowledge to deal with the complexities of hurricane prediction and impact on the Gulf Coast. However,

each of these individuals was working at their respective universities on various specific problems without an overall guiding research umbrella. It was at this meeting that I proposed that we form the Severe Storm Prediction, Education and Evacuation from Disasters (SSPEED) Center to begin to address and evaluate this important problem in the Gulf Coast region.

The SSPEED Center was created to address the need for severe storm research and advanced surge prediction at the Texas Gulf Coast, radar-based rainfall and flood warning systems that combine inland flooding and coastal surge, educational programs that link science with public evacuation and disaster response, and the analysis of infrastructure risk and methods for recovery. The proposal was submitted to the Texas legislature and approved by the governor in May 2007.

On September 13, 2008 Hurricane Ike struck the Texas coastline. As one of the most destructive hurricanes in US history, Hurricane Ike was a devastating event for the region, particularly the Houston/Galveston area. Major damage from wind, surge, and flooding occurred and many areas, including the coastal islands Galveston and Bolivar, have yet to recover. The storm caused $25 billion in damages and dozens of deaths. However, if the hurricane had made landfall 30–50 miles further down the coast, the devastation would have been remarkable. Some research has estimated costs as high as $100 billion and hundreds of fatalities. Hurricane Ike highlighted how exposed and vulnerable the Houston/Galveston area is in the event of a major storm. The decisions that will be made to address this vulnerability will shape the landscape of the Gulf Coast for decades to come in terms of economy, social structure, and environmental policy.

This book was designed to be read by the concerned citizen, policy maker, or urban planner. It bridges the gap between the physical science, social science, and policy in hurricane preparedness and planning to present a brief overview of the premier research being conducted in the wake of Hurricane Ike. Chapters 1–3 introduce the history, characteristics, and meteorology of severe storms with an emphasis on Hurricane Ike. Chapters 4 and 5 address innovative modeling techniques for severe storms, both for inland flooding and storm surge prediction. Chapter 6 addresses the resiliency of coastal communities through social vulnerability mapping. Chapters 7 and 8 primarily deal with communications with the public, specifically emergency management, and evacuation planning. Chapter 9 discusses regional bridge infrastructure and its impacts on transportation surrounding major events. Chapter 10 addresses some of the critical infrastructures impacted by Hurricane Ike and makes some assumptions about impacts of future storms on those critical infrastructures. Finally, chapters 11 and 12 address land use and future sustainability of the Houston/Galvston area.

# Acknowledgments

The authors would like to acknowledge the Houston Endowment and the Texas Medical Center for their generous support of the SSPEED Center as it relates to severe storm prediction and impacts along the Gulf Coast. This book could not have been completed without a group of very devoted associates. We would like to thank Antonia Sebastian, Erin Baker, and Courtney Ray for their contributions to the organization, editing, and review of the book; Garrett Dolan and Alex Wagner for contributions to text organization; and Bryan Carlile and Rik Hovinga for their contributions to figures throughout the book.

We would also like to acknowledge the following people and organizations that contributed time or research funding to the projects presented in the book:

- HNTB
- The National Science Foundation (Grants No. 0901605, 0928926, CMS-0346673)
- The National Oceanographic and Atmospheric Administration (Grant NA08NOS4190458)
- The Texas Department of Rural Affairs
- The Texas Department of Transportation
- Sullivan Brothers Builders and its owners William, Todd, and John Sullivan
- Christy Wilhite—Chief Transportation Planner—and Dmitry Messen—Socioeconomic Program Manager—Houston-Galveston Area Council
- Johnny Cronin, Victoria Herrin, and Elizabeth Winston Jones, of Houston Wilderness

Lessons from Hurricane Ike

# An Introduction to Gulf Coast Severe Storms and Hurricanes

## Philip B. Bedient and Antonia Sebastian

### Introduction

As populations move toward coastal regions in ever increasing numbers, the impacts and associated costs of severe storms on coastal counties increase exponentially. A severe storm is usually considered to be a tropical storm or hurricane, but may also be a severely damaging hail storm, tornado or thunderstorm (Doswell 1978). Each summer, a number of these storms enter the Gulf of Mexico and gain intensity as they hover over its warm waters (fig. 1.1). These storms have the potential to claim hundreds of lives and inflict tens of billions of dollars in damages. The recovery times in impacted areas vary significantly depending on the severity and location of the storm. In some cases, several years are required for infrastructure to fully recover from a severe storm or hurricane.

In 2008, 87 million people were living in US coastal counties, an 84 percent increase from 1960. The Gulf Coast region stretches from the southern tip of Florida to the southern tip of Texas, encompassing 56 counties and more than 5.8 million housing units (Wilson 2010). The region is home to over 15 million inhabitants and is among the most vulnerable areas to severe storms and hurricanes in the United States. In addition, the two largest petrochemical facilities in the world, the two largest refining areas in the United States, and over 4000 oil rig structures are located in this region. In order to fully comprehend and accurately analyze the risks presently faced by the Gulf Coast, an understanding of the mechanics and history of severe storms in the region is necessary, and is covered in detail in this introduction.

Figure 1.1. Measuring over 445 miles across, Hurricane Rita heads towards the Texas/Louisiana border. Rita caused a mass evacuation in Texas and over 2.5 million people left Houston in a matter of 24 hours. Photo courtesy NASA.

**Major Hurricane History**
Data from 1949 in the Pacific, from 1851 in the Atlantic

## Tropical Cyclone Development

North Atlantic tropical cyclones form off the coast of West Africa, over the tropical oceans between 5 and 20 degrees north or south latitude, and in the Gulf of Mexico. Traveling westward, tropical cyclones may make landfall along the coast of the eastern United States, Caribbean, and Central America (fig. 1.2). Those that make landfall in the Caribbean or over southern Florida often have enough strength to cross land and enter the Gulf of Mexico. Despite being severely weakened, these storms can strengthen across the Gulf of Mexico. The warm waters, high humidity, and low wind shear present in the Gulf during summer allow hurricanes to rebuild and gain strength, leaving coastal counties at a great risk for hurricane landfall.

With extreme amounts of rainfall and winds that can exceed 186 mph, these storms are among the most destructive on earth (Bedient et al. 2008).

Figure 1.2. Atlantic Basin tropical cyclone activity 1851–2005. Tropical cyclones that became major hurricanes are colored in yellow. Photo courtesy NOAA.

Severe tropical cyclones have different names depending upon their location (i.e. typhoon, cyclone, baguio); the North American term, used in this discussion, is hurricane. Hurricanes require water temperatures of greater than 80 degrees Fahrenheit at depths up to 50 meters below the surface and warm and moisture-laden air, which possesses an enormous capacity for heat energy. Under these conditions, the condensation from vapor to water sustains hurricane development. An in-depth discussion of the components and mechanics of hurricane formation can be found in Chapter 3.

**Tropical Storms**

Each year, tropical storms (wind speeds 39–73 mph) wreak major damage on coastal communities. Slower wind speeds reduce the risk of damage caused by storm surge and tornados, but increase the risk of flooding because these storms typically have slower forward movement. The most famous tropical storm to make landfall on the Gulf Coast was Tropical Storm Allison. Allison made landfall on the afternoon of June 5, 2001 in Freeport, Texas. The storm quickly weakened into a tropical depression, but continued circulating over the Houston/Galveston Area for four days, dumping incredible amounts of rain-

fall on downtown Houston and the Texas Medical Center. Much of the area received more than 24 inches of rain in nine hours and flooding from the storm caused 41 deaths and more than $5 billion in damages. In parts of downtown Houston and at the Texas Medical Center, three major facilities sustained $300–$400 million in damages each, and suffered power outages during the event.

Although much of the region was unprepared for the event and unaware of their flood risk, the Texas Medical Center had a flood warning system in place. The Rice/TMC Flood Alert (FAS) System developed in 1997 was successful during Tropical Storm Allison, providing advanced warnings for TMC and allowing for some facilities (especially Texas Children's Hospital) to lock down before the event. Tropical Storm Allison and the FAS System are further discussed in Chapter 4.

In the aftermath of Tropical Storm Allison, the Federal Emergency Management Agency (FEMA) and the Harris County Flood Control District (HCFCD) conducted a joint hydrologic study. The goal of the Tropical Storm Allison Recovery Project (TSARP) was to produce more accurate Flood Insurance Rate Maps and delineate flood hazard areas (http://www.tsarp.org/index.asp). This can help assist local recovery and prepare for future events by identifying areas of high risk of flooding and educating the community. To date, Tropical Storm Allison is the costliest and deadliest tropical storm to make landfall in the United States.

## The Galveston Hurricane of 1900

In 1900, the deadliest storm to ever make landfall in the United States destroyed Galveston Island. More than 8000 people died, 8–15 feet of water covered the island, and damages amounted to $30 million (not adjusted for inflation). This storm changed not only the history of the island, but the economy and population of Texas, because most of the shipping activity was moved to the Port of Houston after the event.

After the Galveston Hurricane of 1900, the city made a decision to protect itself from severe storm surge that is often associated with hurricanes. Storm surge is the volume of water pushed ahead of a tropical cyclone as it makes landfall. In one of the greatest engineering feats of its time, Galveston raised the elevation of the entire city, block by block, a project that was completed in 1910. In addition, a 17-foot high, 27-foot wide, and 3-mile long granite boulder seawall was erected along the southern coastline to shield downtown Galveston (fig. 1.3). This landmark was later expanded to 10 miles long in 1962. The hard work and

capital that went into building the seawall and raising the city, though extensive, proved worthwhile. The city was able to successfully withstand landfalling hurricanes in 1915, 1932, 1949, 1983, and most recently, in 2008 with the arrival of Hurricane Ike.

The City of Houston began to experience rapid economic growth after the discovery of oil in 1901 at Spindletop in Beaumont, Texas and many Galveston citizens were motivated to migrate northward after the storm. Once the Houston Ship Channel was dredged in 1909, Galveston lost all hope of reestablishing itself as the premier commercial port on the western Gulf and evolved into a popular beach resort for Texas and much of the central United States.

The track of the Galveston Hurricane of 1900 contributed to the development of strong winds and high storm surge. The hurricane developed into a tropical storm in the central Atlantic, remaining at tropical storm strength until after it entered the Gulf of Mexico. Here, the shallow, warm waters and the atmospheric conditions were favorable for further development. The storm upgraded to a Category 3 hurricane (wind speeds 111–130 mph) by the time it reached the central Gulf. The bathymetry near Galveston helped to exacerbate surge, severely inundating coastal communities. Storm surge, in particular, con-

Figure 1.3. In memory of the victims lost at sea during the Galveston Hurricane of 1900, the statue, "Place of Remembrance," represents the suffering of the victims and the strength of the survivors who stayed to rebuild the city. Photo courtesy Jocelyn Augustino/FEMA.

tributed to the large death toll during the storm of 1900. Hurricane Ike took an eerily similar path to that of the Galveston Hurricane of 1900 and likewise featured devastating storm surge (fig. 1.4).

## Florida Hurricanes

Between 1950 and 1990, the number of hurricanes in the Atlantic varied from 3 to 12 per year, followed by the 1990s during which there was an increase in hurricane activ-

ity in the North Atlantic Basin relative to both frequency and intensity of named storms. In August 1992, Hurricane Andrew made landfall along the southeastern Florida coast with sustained wind speeds as high as 167 mph (Rappaport 2005b), making it a Category 5. Andrew broke hurricane records with a minimum central pressure of 922 millibar measured just before landfall, the third lowest pressure measured for an Atlantic hurricane at that time. In Florida, Hurricane Andrew caused a 23-foot storm surge, 23 deaths, and $26.5 billion in dam-

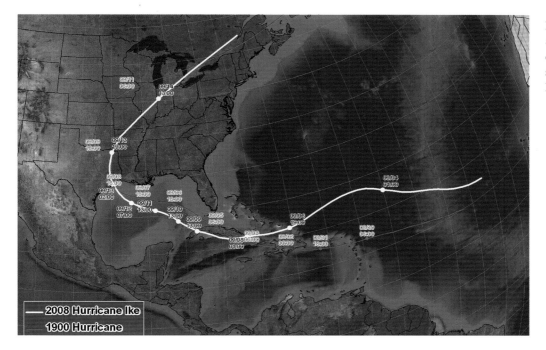

Figure 1.4. Hurricane Ike and the Galveston Hurricane of 1900 took eerily similar paths. Courtesy Rice University Archives.

ages, making it the costliest hurricane to make landfall in the United States during the twentieth century (Rappaport 2005b). The frequency of severe storm events continued to increase after 1992 in the United States and Caribbean Islands and overall hurricane activity doubled between 1995 and 2001 as compared to the previous 23 years (1971–94).

The numbers of Category 4 and 5 storms between 1990 and 2004 also doubled in comparison with the preceding 15 years, and in 2004 some of the most costly hurricanes made landfall in the United States. Four major hurricanes made landfall in Florida in 2004: Charley, Frances, Ivan, and Jeanne. In August, Hurricane Charley made landfall in Fort Myers, Florida, with wind speeds of 150 mph (Pasch et al. 2005) and was classified as a Category 4 hurricane. In the United States, Charley left ten dead and $15 billion in damages, making it the second costliest hurricane at the time. In September, Hurricane Ivan made landfall in Alabama as a Category 3, causing heavy rain and tornadoes before moving back out over the Atlantic, crossing south Florida and making landfall a second time along the Gulf Coast as a tropical depression. Ivan caused storm surge and heavy rain across 15

**Table 1.1 Ten deadliest Atlantic hurricanes making landfall on the mainland US 1900–2009 (Blake et al. 2007).**

| Rank | Name | Year | Category | Deaths |
|---|---|---|---|---|
| 1 | TX (Galveston) | 1900 | 4 | 8000 |
| 2 | FL (SE/Lake Okeechobee) | 1928 | 4 | 2500 |
| 3 | Katrina | 2005 | 3 | 1833 |
| 4 | Audrey | 1957 | 4 | 416 |
| 5 | FL (Keys) | 1935 | 5 | 408 |
| 6 | FL (Miami)/MS/AL/Pensacola | 1926 | 4 | 372 |
| 7 | LA (Grand Isle) | 1909 | 3 | 350 |
| 8 | FL (Keys)/S TX | 1919 | 4 | 287 |
| 9 | LA (New Orleans) | 1915 | 4 | 275 |
| 10 | TX (Galveston) | 1915 | 4 | 275 |

**Table 1.2 Ten costliest Atlantic hurricanes 1900–2009, not adjusted for inflation (Blake et al. 2007).**

| Rank | Name | Year | Category | Cost(billions) |
|---|---|---|---|---|
| 1 | Katrina | 2005 | 3 | $ 81.0 |
| 2 | Andrew | 1992 | 5 | $ 26.5 |
| 3 | Ike | 2008 | 2 | $ 24.9 |
| 4 | Wilma | 2005 | 3 | $ 20.6 |
| 5 | Ivan | 2004 | 3 | $ 14.2 |
| 6 | Charley | 2004 | 4 | $ 15.0 |
| 7 | Rita | 2005 | 3 | $ 10.0 |
| 8 | Frances | 2004 | 2 | $ 8.9 |
| 9 | Hugo | 1989 | 4 | $ 7.0 |
| 10 | Jeanne | 2004 | 3 | $ 6.9 |

states, killing 25 in the United States, and leaving $14.2 billion in damages (Stewart 2005).

The trend of damaging and deadly storms continued into the 2005 hurricane season and for the first time in history, two Category 5 hurricanes entered the Gulf of Mexico during the same season. During August and September, hurricanes Katrina and Rita struck the coasts of Louisiana and Texas. These two hurricanes were unprecedented in recent US history. Hurricane Katrina caused intense storm surge across state lines and breached the levees protecting New Orleans, Louisiana, destroying the city. Hurricane Rita, despite making landfall along the Texas-Louisiana border, a relatively unpopulated region, resulted in thousands of panicked Gulf Coast residents stranded along Texas highways. Tables 1.1 and 1.2 summarize hurricane statistics by depths and damages since 1900.

**Hurricane Katrina**

On August 29, 2005, Hurricane Katrina made landfall as a Category 3 storm just east of New Orleans, Louisiana. Katrina caused 1833 deaths and resulted in $81 billion in damages making it the third deadliest and the costliest storm to make landfall in the United States (Knabb et al. 2006a).

Katrina had wind speeds of 127 mph at landfall and also set other records, such as the third lowest central pressure at landfall (920 millibar) and the sixth lowest central pressure on record (902 millibar). However, the most notable legacy Hurricane Katrina left in its wake was the destruction of the levee system surrounding New Orleans and the devastation and flooding that ensued. Many New Orleans residents never returned after the storm and the city suffered significant population loss.

On August 23, 2005, Hurricane Katrina developed as a tropical depression near the Bahamas and on August 25, Katrina made landfall at the southern coast of Florida as a weak hurricane. Katrina then began gaining strength as it entered the Gulf of Mexico. On August 28, Katrina reached its maximum strength as a Category 5 hurricane with sustained winds of 173 mph (fig. 1.5). One day later, on August 29, Katrina made landfall as a Category 3 in New Orleans, Louisiana.

New Orleans lies 6–20 feet below sea level. The city is surrounded by a series of levees protecting it from Lake Pontchartrain and the Mississippi River, both of which empty into the Gulf of Mexico (fig. 1.6). On August 28, 2005 one day before making landfall, Katrina was forecasted to cause storm surge as high as 28 feet, breaching even the highest levee in New Orleans. City and federal officials feared the worst and ordered the first-ever mandatory evacuation of the city.

These fears were confirmed when 53 levees were breached and the 40 Arpent Canal Levee failed as Katrina made landfall. Approximately 131 billion gallons of water flowed into the City of New Orleans with nowhere to go (fig. 1.7). The water eventually subsided, with the help of sandbag damming and pumping efforts by the Army Corps of Engineers (fig. 1.8). Despite these efforts, the city of New Orleans would be forever changed, most of its homes washed away with the surge. Countless people evacuated to nearby parishes and states, and schools across the United States opened their doors to students from New Orleans and other affected areas (fig. 1.9). Many families were forced to stay out of New Orleans and parts of Louisiana for months after the storm and many never returned.

Hurricane Katrina illustrated the degree of social disruption that could occur as a result of a hurricane. It served to accentuate the vulnerability of certain Gulf Coast communities, a point that would be underscored just three weeks later when Hurricane Rita headed toward Texas (Bedient et al. 2008).

Figure 1.5. Hurricane Katrina approaching Louisiana in 2005. Katrina, the most devastating natural disaster in recent US history, took the lives of over 1800 people and caused over $133 billion in damages. Photo courtesy NASA.

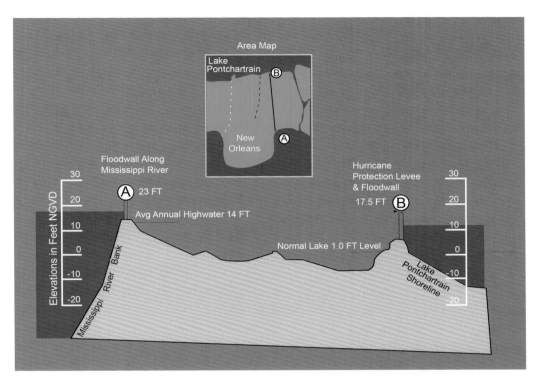

Figure 1.6. The City of New Orleans is below sea level and bordered by the Mississippi River and Lake Ponchartrain. Modified from US Army Corps of Engineers.

Figure 1.7. Following Hurricane Katrina, a broken levee along the Industrial Canal allows water to flow into the City of New Orleans. Levees surrounding the city broke, dumping over 131 billion gallons of water into an area of homes and businesses. Photo courtesy FEMA.

Figure 1.8. A Blackhawk helicopter drops sandbags on top of a broken levee after Hurricane Katrina. Levees surrounding the city were originally put in as part of the Flood Control Act of 1965. Photo courtesy Jocelyn Augustino/FEMA.

Figure 1.9. Three days after Katrina made landfall, five of the 1355 buses used to evacuate New Orleans citizens head to Houston, Texas. Houston, along with other parts of Louisiana, Mississippi, and Alabama became home to thousands of displaced residents. Photo courtesy Marty Bahamonde/FEMA.

Table 1.3 Mainland US hurricanes by Saffir–Simpson category, 1851–2006.

| Decade | 1 | 2 | 3 | 4 | 5 | All (1–5) | Major (3–5) |
|---|---|---|---|---|---|---|---|
| 1851–1860 | 7 | 5 | 5 | 1 | 0 | 18 | 6 |
| 1861–1870 | 8 | 6 | 1 | 0 | 0 | 15 | 1 |
| 1871–1880 | 7 | 6 | 7 | 0 | 0 | 20 | 7 |
| 1881–1890 | 8 | 9 | 4 | 1 | 0 | 22 | 5 |
| 1891–1900 | 8 | 5 | 5 | 3 | 0 | 21 | 8 |
| 1901–1910 | 10 | 4 | 4 | 0 | 0 | 18 | 4 |
| 1911–1920 | 10 | 4 | 4 | 3 | 0 | 21 | 7 |
| 1921–1930 | 5 | 3 | 3 | 2 | 0 | 13 | 5 |
| 1931–1940 | 4 | 7 | 6 | 1 | 1 | 19 | 8 |
| 1941–1950 | 8 | 6 | 9 | 1 | 0 | 24 | 10 |
| 1951–1960 | 8 | 1 | 6 | 3 | 0 | 18 | 9 |
| 1961–1970 | 3 | 5 | 4 | 1 | 1 | 14 | 6 |
| 1971–1980 | 6 | 2 | 4 | 0 | 0 | 12 | 4 |
| 1981–1990 | 9 | 2 | 3 | 1 | 0 | 15 | 4 |
| 1991–2000 | 3 | 6 | 4 | 0 | 1 | 14 | 5 |
| 2001–2010 | 17 | 8 | 9 | 9 | 2 | 45 | 20 |
| | | | | | | | |
| 1851–2006 | 110 | 73 | 75 | 18 | 3 | 279 | 96 |
| Average per decade | 6.9 | 4.6 | 4.7 | 1.1 | 0.2 | 17.4 | 6.0 |

## Hurricane Rita

On September 23, 2005, Hurricane Rita made landfall near the Texas-Louisiana border as a Category 3 storm. Another record breaking storm, Hurricane Rita had the fourth lowest central pressure on record (895 millibar) (Knabb et al. 2006b), replacing Hurricane Katrina. With much of the Gulf Coast region still reeling after Katrina, Hurricane Rita triggered the largest urban evacuation in US history. More than 2.5 million people left the Houston/Galveston Area in a 48-hour period. On September 22, one day before landfall, evacuation contra-flow lanes were opened on I-45 from Houston towards Dallas, on I-10 from Houston towards San Antonio, and on Highway 290 from Houston towards Austin. Trips between cities that normally take just a few hours ranged from 10 to 36 hours and had Rita made landfall further south, many would not have reached their destination in time.

Figure 1.10. Traffic on I-45 as residents in the Houston/Galveston area evacuated before Hurricane Rita. Photo courtesy Ed Edahl/FEMA.

Hurricane Rita began developing as a tropical depression east of the island of Grand Turk on September 17, 2005. Upon entering the Gulf of Mexico, Hurricane Rita gained strength and began intensifying until, on September 21, Rita reached Category 5 status with wind speeds of 159 mph. Only 6 hours later, Hurricane Rita reached its maximum intensity with 180 mph sustained winds and a minimum pressure of 895 millibar. On September 23, Hurricane Rita made landfall between Sabine Pass, Texas and Johnson's Bayou, Louisiana as a Category 3 with maximum sustained winds of 115 mph and a minimum central pressure of 897 millibar.

Although Hurricane Rita made landfall in a rural area, flooding caused seven deaths and damages amounting to $10 billion. Nine counties in the Beaumont/Port Arthur region were declared disaster areas by Governor Rick Perry. While unfortunate and certainly not insignificant, the damage attributed to Rita would have been much worse had it made landfall in the Houston/Galveston Area. However, as Rita barreled across the Gulf of Mexico intense panic ensued across Gulf Coast region and quickly the highways were packed (fig. 1.10). The highways heading toward Dallas, Austin and San Antonio were at a standstill; many ran out gas and some even died of heat exhaustion.

The initial forecasted path of Rita

towards Houston and the recentness of the disaster caused by Hurricane Katrina, left Texas Gulf Coast residents feeling extremely vulnerable. However, the area was ill-prepared for the panic that occurred. Because of this, Hurricane Rita is one of the most memorable hurricanes in recent history, well-known for the intense panic and disorganization that it caused across the area. For politicians and public policy, Hurricane Rita stood as a glaring example of a failure in preparedness. Since Rita, the area has taken gigantic steps towards disaster preparedness and against the "Rita effect," the havoc that ensued in the wake of the storm.

## Conclusion

Although the past decade has seen increased hurricane activity, similar periods of heightened activity have also been seen. Table 1.3 summarizes hurricane activity by decade since 1851.

Weather patterns during periods of heightened activity are often characterized by sea surface temperatures. More active seasons also occur when there is a subtropical ridge across the central and eastern North Atlantic, and an African Easterly Jet that is favorable to developing and intensifying tropical disturbances moving westward from the east coast of Africa (Bedient et al. 2008). Because of

differences in meteorological technologies, no significant increasing trend has been established between the active phases of the oscillation. Although increased atmospheric temperature from greenhouse gases could theoretically increase hurricane intensity, no direct link has been established by meteorologists. Interestingly, 2006 was a particularly quiet hurricane season with very few storms reported (Bedient et al. 2008).

The relative quiet of the 2006 hurricane was contrasted in 2008 when the costliest and most damaging storm to hit the Texas coast since the Galveston Hurricane of 1900 made landfall at Galveston, Texas on September 13, 2008. Hurricane Ike made landfall as a Category 2, but its size resembled that of a Category 4 hurricane. The resulting high water and wind caused over $24.9 billion damages and claimed at least 103 lives (NOAA 2010). Its effects were widespread due to its large size and intensity. The following chapter covers the formation and landfall of Hurricane Ike in detail. The remainder of the book discusses methods for predicting severe storms and hurricanes, vulnerabilities of society and infrastructure, disaster response and mitigation, and future planning in the wake of Hurricane Ike.

# 2

# Hurricane Ike

## Philip B. Bedient and Antonia Sebastian

### Introduction

The arrival of Hurricane Ike on the Texas coast on September 13, 2008 marked the beginning of a re-evaluation of Gulf Coast hurricanes, both in terms of their perceived potential damage and in the ways that communities choose to protect themselves. Ike was the most destructive storm to make landfall on the Texas coast since the Galveston Hurricane of 1900 and with $24.9 billion in damages, Ike was the third costliest storm in US history (Berg 2010).

At its peak intensity, Ike was characterized as a Category 4 hurricane with sustained winds reaching 145 mph and a minimum central pressure of 935 mbar (fig. 2.1). The wind field of Hurricane Ike spanned 450 miles at landfall, covering most of Texas and parts of Louisiana. In Texas, wind damage caused 2.6 million power outages and many people were without power until early October and some, even longer. The eye of Ike passed just north of downtown Houston and on the morning of September 13, the

streets of downtown were littered with broken glass from the skyscrapers, blown out by the pressure of the storm and high winds (fig. 2.2). Trees across the city were uprooted, roads were blocked, and homes and cars crushed.

Hurricane Ike brought comparatively little rainfall to the Houston/Galveston Area, but storm surge caused significant flooding along the barrier Peninsula from Seaside Beach to Bolivar Island. It was determined that the storm surge was 17.8 feet on Bolivar Peninsula. One high watermark, collected by FEMA, was at 17.5 feet, located in Chambers County, approximately 10 miles inland (Berg 2010). After the water receded, the destruction on Bolivar Peninsula was nearly complete (fig. 2.3a). On Galveston Island, many homes were flooded and beachside houses were washed away. Debris was piled up all over the island (fig. 2.3b). Along the Galveston Sea Wall, no beach sand was left and on the island, many live oak trees were permanently damaged by salt water and could not be saved.

As Ike traveled north across the

Figure 2.1 Prominent bands of Hurricane Ike as seen from the International Space Station. Photo courtesy NASA.

Figure 2.2. Broken window damage on the JPMorgan Chase building downtown Houston after Hurricane Ike. Houston police closed off streets surrounding the building due to extensive damage. Photo courtesy James Nielsen/Houston Chronicle.

Table 2.1. Hurricane Ike timeline.

| Date and Time (2008) | Event |
| --- | --- |
| August 28 | Ike originates off the coast of West Africa as a tropical wave |
| September 1 | Ike develops into a tropical storm west of the Cape Verde Islands |
| September 3 | Ike is classified first as a hurricane and, within 3 hours of first reaching hurricane strength, is reclassified as a major hurricane with maximum wind speeds of 115 mph |
| September 4 | Ike is reclassified as a Category 4 hurricane with maximum wind speeds of 145 mph |
| September 7 | Ike makes landfall over the Turks and Caicos Islands |
| September 8–9 | Ike makes landfall over Cuba and enters the Gulf of Mexico as a Category 2 hurricane |
| September 10 | Ike continues on its track towards Galveston Island as a large Category 2 hurricane with a pressure of 944 millibar |
| September 11 | Ike is measured at 450 miles wide; NOAA issues a Hurricane Warning for the area between Morgan City, Louisiana and Baffin Bay, Texas |
| September 12 | The storm surge effects of Hurricane Ike are felt along the Texas coast 24 hours before the storm makes landfall |
| 2:10 am CDT; September 13 | Hurricane Ike makes landfall at Galveston, Texas as a Category 2 storm with winds of 110 mph and a central pressure of 950 mbar |
| 4:00 am CDT; September 13 | Hurricane Ike makes landfall near Baytown, Texas |
| 1:00 am CDT; September 13 | Ike is downgraded to a tropical depression as it passes 100 miles east of Dallas, Texas |
| Late October, 2008 | Most energy customers with the ability to be connected to the power grid have power returned to their homes |

Figure 2.3a. Hurricane Ike's ferocious storm surge leaves Bolivar Peninsula in shambles. The peninsula was one of the hardest hit areas, leaving the island with permanent beach erosion and destroying many homes. Photo © 2011 Bryan Carlile, Beck Geodetix.

Figure 2.3b. A debris pile, including a boat, sits on the front lawn of Galveston Island homes destroyed during Hurricane Ike. Photo courtesy Robert Kaufmann/FEMA.

continental United States it caused flooding in 11 states and resulted in over 20 deaths. As of May 3, 2010, there were still 23 people unaccounted for, 16 of which were Galveston Island residents. The social, economic, and environmental impacts of Hurricane Ike on the region remain to this day because many of the people and homes and much of the vegetation of the coast is still gone. Since Ike, studies have been completed, ranging from an evaluation of the preparedness of the region, to hurricane modeling including varied geographic scenarios of Ike's landfall. In subsequent chapters, these studies will be discussed in depth.

## Hurricane Development

Hurricane Ike began as a tropical wave formation off the coast of West Africa on August 28, 2008. The timeline associated with Ike is shown in table 2.1. As the wave moved north, northwest across the Atlantic, it intensified and developed enough convective organization to be classified as a Tropical Depression. On September 1, 2008, 5:00 p.m. EDT, Ike became a Tropical Storm. Ike moved across the Atlantic Ocean and wind shear north of the storm began to weaken the storm slowing it to a forward speed of 18 mph.

On the morning of September 3, an eye began to develop and by that afternoon, Ike had strengthened enough to become a weak hurricane. Just three hours later, Ike was upgraded to a "major storm" with winds of 115 mph. After another three hours, Ike was upgraded to a Category 4 storm with winds reaching 135 mph and an estimated central pressure of 948 millibar (Berg 2010). Hurricane Ike achieved its peak intensity with winds of 145 mph and an estimated central pressure of 935 millibar on the morning of September 4.

During the afternoon of September 4, Ike dropped in intensity, only to be followed by a re-intensification during the evening. Warming on September 5 reduced Ike to a Category 3 storm and increased wind shear on September 6 resulted in further downgrading to a Category 2. However, six hours later Hurricane Ike intensified and was again upgraded to a Category 4. Hurricane Ike maintained its intensity as it made landfall on the Turks and Caicos Islands on the morning of September 7 and in the evening, Ike made landfall on the north coast of Cuba as a Category 3 hurricane (Berg 2010).

On the afternoon of September 9, Hurricane Ike entered the Gulf of Mexico and the next day as Ike barreled towards Texas, the intensity of the storm increased when its central pressure dropped from 963 millibar to 944 millibar. On September 11, despite only being a Category 2 storm with winds of 110 mph, the wind field of Ike was atypically large, 450 miles in diameter. By early morning the next day, the effects of Hurricane Ike were felt along the Texas coast as waves began hammering against the Galveston sea wall.

On September 13, at 2:10 a.m. CDT, Hurricane Ike made landfall at Galveston, Texas as a Category 2 storm with sustained winds of 110 mph and a central pressure of 950 millibar. At 4:00 a.m. CDT, Ike had crossed Galveston and was near Baytown, Texas and by 1:00 p.m. CDT on September 13, Ike was downgraded to a tropical depres-

sion as it passed 100 miles east of Dallas, Texas (Berg 2010). As a tropical depression, Ike made its way across the United States and left flooding and destruction in its wake (fig. 2.4).

**Storm Surge**

On Friday September 12, 24 hours ahead of landfall, storm surge associated with Hurricane Ike was felt along the Gulf Coast (fig. 2.5). By Friday evening, surge levels had neared 10 feet on Galveston Island, attaining maximum depths of 15–20 feet on Bolivar Peninsula (Berg 2010). When Hurri-

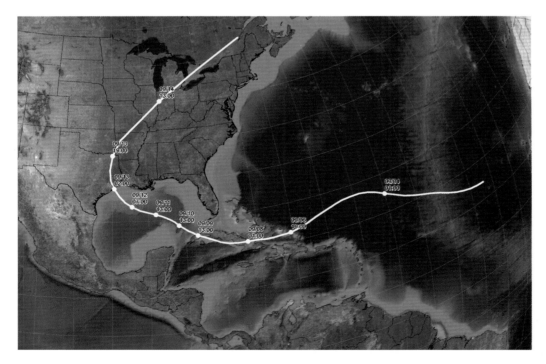

Figure 2.4. The path of Hurricane Ike. Courtesy Rice University Archives.

# Hurricane Ike Inundation Depth

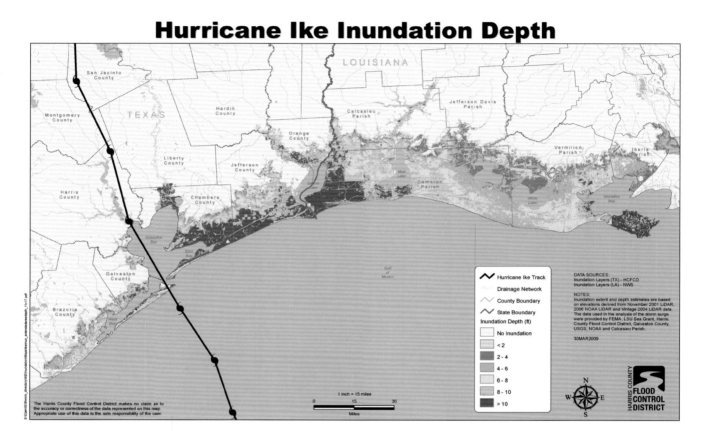

Figure 2.5. Storm surge inundation along the Gulf Coast caused by Hurricane Ike. Photo courtesy Harris County Flood Control District (HCFCD).

cane Ike made landfall, its wind speeds were below 110 mph. However, storm surge coupled with high tide and the size of Ike's wind field resulted in incredible damage from storm surge and wind. Even inland, along Galveston Bay, homes, boats, and docks were destroyed. On Galveston Island, flooding began on the bay side of the island where the land surface was unprotected by the 17-foot high seawall that stands between the front of the island and the waters of the Gulf of Mexico.

On Bolivar Peninsula, surge ripped through structures and inundated the peninsula, leaving massive piles of rubble after receding.

Storm surge also greatly affected the natural environments of the Texas Gulf Coast. Saltwater intrusion proved to be a substantial issue, transforming brackish estuaries into saltwater estuaries, as well as killing thousands of live oak trees on Galveston Island, some of which had survived the Galveston Hurricane of 1900. As of October 1, 2008 more than 200 pollution incidents had been reported of which the long-term effects are still unknown. Approximately 60 percent of the oyster beds in Galveston Bay were destroyed and another 50–60 percent were estimated to have been impacted by sedimentation on reefs. Erosion was also a serious problem during and after Ike. Approximately 300 feet of beach was lost to the storm. The detailed impacts of the storm are discussed in later chapters.

## Power Loss

Aside from storm surge, the other legacy of Hurricane Ike was power loss on a scale never before witnessed in the Houston/Galveston Area, much less the state of Texas. Approximately 2.6 million residents were left without electricity during the largest power outage in Texas history; significantly larger than the 719,000 and 750,000 people left without power after hurricanes Rita (2005) and Alicia (1983), respectively (CenterPoint 2010).

The largest power provider in the area, CenterPoint Energy, reported an initial loss of power for over 90 percent of its users. To fix the problem, CenterPoint alone summoned 5000 employees and 10,000 line workers, many from out of state, to deal with the power loss (fig. 2.6). They replaced

6400 of the 1 million wooden distribution poles in the area. Regional power distributors continued to work to restore electricity. Initial losses to the company as a result of the storm were estimated to be as high as $500 million (Palmeri 2008). Although many areas had power restored within a few days or weeks, some areas were without power months after the storm hit. This great loss of power was largely due to falling trees resulting from strong winds during the storm.

## Associated Costs of the Storm

Many of the lasting effects of Hurricane Ike are easily quantified and serve to underscore the severity of the storm. With an estimated total damage of $24.9 billion, Ike is the third costliest storm in US history, trailing hurricanes Katrina (2005) and Andrew (1992). Perhaps the most striking losses were observed on Bolivar Peninsula, where the county engineer estimated that over 3000 structures were destroyed and an additional 140 structures damaged. While this number is a small percentage of the total number of structures damaged during Ike, the community on Bolivar was relatively small and emigration from the area due to home loss resulted in a 70 percent loss in tax revenue to the area (FEMA 2008).

Figure 2.6. To help mitigate power outages, Entergy sends in power trucks from Port Fourchon, Louisiana to begin repair on damaged power lines. Over 2.15 million CenterPoint Energy customers were without power, resulting in the largest power outage in Texas history. Photo courtesy Entergy Services, Inc..

Across the region, residential costs resulting from wind and flood damage totaled around $3.4 billion. Of the homes damaged, approximately 27 percent of the wind-damaged homes and 61 percent of the flood-damaged homes were uninsured. By December 1, 2008, FEMA had paid out over $20 million in housing assistance to over 100,000 applicants. Although 92 percent of the residential homes in the Houston Metropolitan Statistical Area were undamaged, more than 150,000 homes were damaged to varying degrees ranging from mi-

nor to complete destruction (fig. 2.7). A summary of those damaged homes is given in table 2.2.

The healthcare, petrochemical, and agricultural industries were among those most affected by the storm. The healthcare industry experienced a $40 million deficit per month, having lost 750,000 square feet of building space. They compensated by scaling back on 3800 jobs. The University of Texas Medical Branch at Galveston incurred damage costs of $710 million. The petrochemical industry estimated damages in the hundreds of millions of

Table 2.2. Summary of homes damaged during Hurricane Ike (FEMA 2008).

| Classification | # of Homes | Description of Damage |
|---|---|---|
| Likely to Rebuild | 114,843 | Minor damage estimated at less than $15,000 |
| Likely to be Abandoned | 2,149 | Based on the assumption that 20% of homes with major damage will be abandoned; also assumes that high-value homes are more likely to be repaired |
| Uncertain | 2,687 | 25% of homes with major damage; dependent on presence of insurance, desire, and cost |

Figure 2.7. An aerial view of what remains in a neighborhood on Bolivar Peninsula. According to NOAA, wind speeds exceeded 110 mph. Photo © 2011 Bryan Carlile, Beck Geodetix.

dollars as a result of saltwater intrusion, and the agricultural industry estimated damages to be approximately $93 million due to substantial damage to livestock, feed stores, fencing, and crops (FEMA 2008).

The State of Texas estimated a total cost of $2.4 billion for waterway transportation maintenance such as erosion mitigation and dredging. Another $131.8 million was estimated to repair affected transportation systems. The City of Galveston reported $1.7 billion in damages to government buildings and wastewater treatment plants (fig. 2.8). Fortunately much of the potential damage to waste disposal locations did

not occur, but the possibility may provide the impetus for preparation for the next storm (FEMA 2008). Damage to critical infrastructures during Hurricane Ike is discussed in chapter 10.

**Looking to the Future**

The severity and location of damage as a result of Hurricane Ike, although significant, has served to underscore the unprecedented loss that would have occurred had this storm made landfall further down the coast or had sustained more powerful winds. As a result, several key areas have

Figure 2.8. US Army Corps of Engineers Greg Lugar of the Critical Public Facilities Team, left, meets with Assistant Fire Chief Jeff Smith at Station 4 in Galveston, Texas after Hurricane Ike. Photo courtesy Jocelyn Augustino/FEMA.

been identified for potential concern: coastal flood insurance, structural v. non-structural flood mitigation, and future development.

The total value of insurance claims made on homes located directly on the coast, coupled with the relatively low cost of flood insurance premiums, has led many to question the viability of federally subsidized flood insurance, questioning whether the premium costs accurately reflect the risk associated with living so near to the coast. Additionally, the cost of flood damage as a result of Hurricane Ike has underscored the need to prepare for potentially more powerful storms, whether through structural or non-structural mitigation. This has quickly become a high profile debate in the Houston/Galveston area. Some proponents of structural mitigation support a proposal to construct a massive dike along the length of Galveston Bay, extending the current sea wall as far as Bolivar Peninsula, barring the intrusion of storm surge. However, opponents are advocating for a managed retreat from certain coastal areas, as well as smaller scale structural projects, discussed in chapter 12.

With upwards of 1 million more people estimated to move into the affected areas in the next few decades, more emphasis is being placed on actively developing the area to reflect real risks posed by hurricanes. It will be necessary to model possible storm surge and inland flooding to identify the most vulnerable coastal areas and to implement more informed development practices.

**Conclusion**

Hurricane Ike was a devastating event for the Gulf Coast region, particularly the Houston/Galveston area. Major damage from wind, storm surge and associated flooding occurred and many areas have yet to recover. Over 2500 homes experienced major damaged with many completely destroyed. Ike was the third costliest storm in US history and its economic impacts continue to be felt, especially on Galveston Island and Bolivar Peninsula. As devastating as Hurricane Ike was, computer analysis indicates that it was not the worst case scenario and winds and storm surge could have been significantly higher in Galveston Bay and near the Houston Ship Channel.

# 3

# A Brief Introduction to the Meteorology of Tropical Cyclones

*Jeffrey Lindner*

## Introduction

Every year on average of 60–100 tropical waves emerge off the west coast of Africa and traverse the tropical Atlantic Ocean, Caribbean Sea, and Gulf of Mexico. Approximately 10–20 of these will develop into tropical cyclones in the Atlantic Basin. Tropical waves originating from Africa are the most common seedlings for tropical cyclone formation, but other types of weather disturbances can lead to their formation as well (fig. 3.1). On a global scale, tropical storm formation has been correlated with several climatic anomalies, including rainfall in West Africa in the prior year, the direction of the winds in the stratosphere, and the El Niño-Southern Oscillation (ENSO) phenomenon.

El Niño is characterized by a warm phase associated with high sea surface temperatures off the coast of Peru, low atmospheric pressure over the eastern Pacific, and increased vertical wind shear over the Atlantic. La Niña, on the other hand, is characterized by a cold phase with low sea surface temperatures in the eastern Pacific, low atmospheric pressure over the western Pacific, and decreased vertical wind shear over the Atlantic (Bedient et al. 2008). As a result of this decrease in vertical wind shear, La Niña often corresponds with increased tropical cyclone activity in the Atlantic Basin. Although the opposite is true for El Niño, powerful storms have still been known to develop while it is in phase, sometimes bringing devastation to US coastal and inland communities.

In the Gulf of Mexico and along the southeastern coast of the United States, stalling frontal boundaries, especially in June and October, may result in a surface low pressure formation that gradually develops into a tropical cyclone. Other less common weather features that develop into tropical cyclones include decaying Mesoscale Convective Systems (MCS), which move off the US mainland or off Central America with weak surface

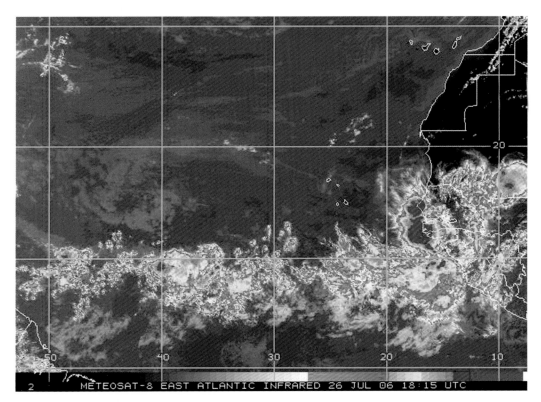

METEOSAT-8 EAST ATLANTIC INFRARED 26 JUL 06 18:15 UTC

Figure 3.1. Weather disturbances off the west coast of Africa move westward with the possibility of developing into tropical cyclones. Photo courtesy NASA.

low pressure systems. Such systems gradually deepen over warm water and become upper level lows that move from east to west within the tropical latitudes. Such lows are typically very slow to transform from a cold core system into a warm core tropical system with thunderstorm activity focused near the center. The process of becoming a warm core tropical system can take several days.

## Tropical Cyclone Formation

The prime hurricane season lasts from June 1 to November 30 when warmer oceans provide the requisite heat and moisture to burgeoning storms. In the Northern Hemisphere, the most likely window for hurricane formation begins in late summer and continues into early fall; however, there exists considerable variability in the annual number of hurricanes in the Atlantic annually. The Atlantic Basin hurricane season peaks in early to mid-September

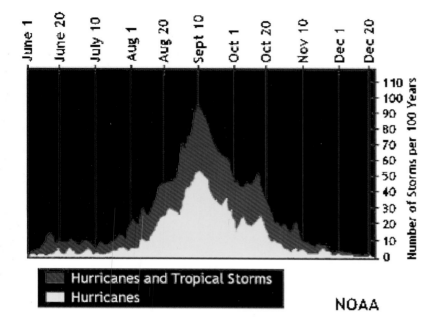

Figure 3.2. The Atlantic hurricane season lasts from June 1 to November 1, peaking in mid-September. Chart courtesy NOAA.

with formation favorable across nearly the entire basin in August and September (fig. 3.2). During June and July, development typically occurs in the western part of the Caribbean Sea, the Gulf of Mexico, and in the region from the Bahamas to North Carolina. In October and November, the most favorable conditions are found in the western Caribbean Sea and the southern Gulf of Mexico.

The first formative stage of a tropical cyclone is a tropical depression, which is classified when an area of disturbed weather closes off a low level circulation, has sustained winds of less than 39 mph, and consists of deep convection near the center of circula-

tion. A tropical storm has sustained winds of 39–73 mph and the system is named. All tropical storms and hurricanes are given proper names in alphabetical order starting with "A" on June 1 and alternating between male and female names. The names cycle every six years (table 3.1). However, some names are retired after significant loss of life and property (NOAA 2010; http://www.nhc.noaa.gov/retirednames.shtml).

When sustained winds reach 74 mph, a tropical storm is upgraded to a hurricane. Hurricanes typically have well defined convection near and surrounding the center with decent banding features. The Saffir-Simpson Hurricane Wind Scale is used to categorize hurricanes based on sustained wind speeds (http://www.nhc.noaa.gov/pdf/sshws_table.pdf). The categories range from Category 1, a hurricane of minimal damage, to Category 5, a hurricane of catastrophic proportions (table 3.2) (NOAA 2010). On average, the potential damage factor rises by a factor of four for each category increase.

Several factors determine the growth and decay of a tropical cyclone. The most important factors are sea surface temperatures (SSTs), vertical wind shear, and atmospheric moisture. For example, tropical cyclones cannot form at the equator because the Coriolis force is weakest at this latitude hindering the rotation of winds.

**Table 3.1. Atlantic hurricane names, 2010–2015.**

| 2010 | 2011 | 2012 | 2013 | 2014 | 2015 |
|------|------|------|------|------|------|
| Alex | Arlene | Alberto | Andrea | Arthur | Ana |
| Bonnie | Bret | Beryl | Barry | Bertha | Bill |
| Colin | Cindy | Chris | Chantal | Cristobal | Claudette |
| Danielle | Don | Debby | Dorian | Dolly | Danny |
| Earl | Emily | Ernesto | Erin | Edouard | Erika |
| Fiona | Franklin | Florence | Fernand | Fay | Fred |
| Gaston | Gert | Gordon | Gabrielle | Gonzalo | Grace |
| Hermine | Harvey | Helene | Humberto | Hanna | Henri |
| Igor | Irene | Isaac | Ingrid | Isaias | Ida |
| Julia | Jose | Joyce | Jerry | Josephine | Joaquin |
| Karl | Katia | Kirk | Karen | Kyle | Kate |
| Lisa | Lee | Leslie | Lorenzo | Laura | Larry |
| Matthew | Maria | Michael | Melissa | Marco | Mindy |
| Nicole | Nate | Nadine | Nestor | Nana | Nicholas |
| Otto | Ophelia | Oscar | Olga | Omar | Odette |
| Paula | Philippe | Patty | Pablo | Paulette | Peter |
| Richard | Rina | Rafael | Rebekah | Rene | Rose |
| Shary | Sean | Sandy | Sebastien | Sally | Sam |
| Tomas | Tammy | Tony | Tanya | Teddy | Teresa |
| Virginie | Vince | Valerie | Van | Vicky | Victor |
| Walter | Whitney | William | Wendy | Wilfred | Wanda |

**Table 3.2. Saffir-Simpson Hurricane Wind Scale.**

| Category | 1-minute sustained wind speed (mph) | Extent of damage | Description of damage |
|----------|------|------|------|
| Tropical Storm | 39–73 | Minor | Some flooding. |
| 1 | 74–95 | Minimal | Very dangerous winds will produce some damage. |
| 2 | 96–110 | Moderate | Extremely dangerous winds will cause extensive damage. |
| 3 | 111–130 | Extensive | Devastating damage will occur. |
| 4 | 131–155 | Extreme | Catastrophic damage will occur. |
| 5 | >155 | Catastrophic | Catastrophic damage will occur. |

Sea surface temperatures conducive to formation are usually at or above the 26–28 degrees Celsius (79–82 degrees Fahrenheit), although in rare instances, tropical cyclones can form over cooler waters. Slow moving tropical systems can also up-well, or bring cooler water from below to the surface, lowering the sea surface temperature below the required threshold to support continued growth. The deeper the warm water, the greater the potential for intensification as experienced in recent hurricane events over the deep swath of warm water in the southeastern Gulf of Mexico, known as the Loop Current.

Tropical cyclones warm the air column through the release of latent heat during intense convection. As this heating occurs, surface pressures lower and cyclonic winds bring inflow towards the developing center (fig. 3.3). This phenomenon is most readily observed under light wind shear when latent heat release is not removed from the air column by strong winds aloft. Tropical cyclones thrive under upper level anticyclones, or high pressure cells, that create light and variable winds aloft. Strong wind shear can cause rapid decay of the inner core of a tropical cyclone and greatly disrupt the circulation pattern. For instance, the mid level center can become detached from the low level center or the entire system can become tilted in the

direction to which the shear is blowing. There are two types of shearing effects: strong low level easterly flow that can force the low level center out ahead of the mid level center, or westerly mid and upper level shear that can pull the top off the tropical cyclone, exposing the low level center.

Without deep convection near or over the center, a tropical cyclone will gradually spin down and weaken, possibly opening up into an open wave instead of a closed, low level circulation. Another important factor is the amount of moisture available for deep convection. Deep convection requires tremendous amounts of moisture,

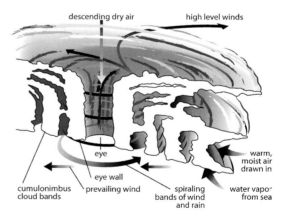

Figure 3.3. Hurricanes rotate in a counter-clockwise direction. At the center of the hurricane, there is a calm area known as the eye of the storm, usually 20–40 miles in diameter. Beyond the eye is the eye wall and a series of rain bands. From Merriam-Webster's Collegiate® Encyclopedia©2000 by Merriam-Webster, Incorporated (www.Merriam-Webster.com) and used with permission.

which in most cases, is readily available over the tropical oceans. Large amounts of moisture are needed in the mid levels of the atmosphere to maintain saturation of an air parcel through the ascent process. Relative humidity in the 50–70 percent range from the 700–500 millibar level is generally sufficient to preclude large amounts of evaporation within updrafts and provide a favorable moist environment for tropical cyclone growth.

**Impacts of Tropical Cyclones**

Tropical cyclones can impact large areas, causing rainfall and inland flooding, damaging winds, storm surge and wave action, and tornadoes. Rainfall and inland flooding are the deadliest aspects of a tropical cyclone. Tropical cyclones are capable of producing incredible amounts of rainfall in short periods of time leading to rapid flash flooding (fig. 3.4). Additionally, such devastating rains can occur well inland and sometimes several days after landfall in regions typically unaccustomed to tropical rains or systems. On average, a tropical cyclone will produce 10–15 inches of rainfall along its track. The most important feature related to the amount of rainfall is how fast the storm is moving, or its forward motion. The slower a tropical system is moving the greater the rainfall over a

given area for a given period of time. Many floods of record and historic flood events are tied to the landfall and subsequent decay of tropical cyclones. Tropical Storm Allison, which made landfall in Freeport, Texas in 2001, stalled over the Houston area causing heavy rainfall and significant flooding in the Texas Medical Center and Downtown (fig. 3.5).

Strong damaging winds are common within tropical cyclones, especially hurricanes. In most cases, the strongest winds are found near the center of the storm surrounding the eye of a well- developed hurricane. The wind field of a hurricane can be very small (Hurricane Charley, 2004) or extremely large (Hurricane Ike, 2008). The size of the wind field and the intensity of the maximum sustained winds are important in determining the potential for damage during landfall. A large wind field can impact a greater area for an extended period of time, whereas a small, intense hurricane can bring devastating damage to a small spatial area.

In 2004, Hurricane Charley made landfall near Punta Gorda, Florida with sustained winds of 150 mph, however this very intense band of winds was concentrated around a small eye and only covered a 20–30 mile area. Near complete devastation of structures occurred in the area. In comparison, Hurricane Ike in 2008

Figure 3.4. Massive amounts of rain produced by Tropical Storm Allison overwhelmed the internal drainage system of the City of Houston causing Buffalo Bayou, which flows through downtown Houston, to overflow its banks. Storm water flowed into the Downtown Houston Tunnel System and into many downtown businesses. Photo courtesy Harris County Flood Control District (HCFCD).

Figure 3.5. Tropical Storm Allison dumped heavy rain on downtown Houston, forcing surrounding highways to be closed. Allison was the costliest and deadliest tropical storm to hit the United States. Photo courtesy Smiley Pool/ Houston Chronicle.

exhibited a massive swath of hurricane force winds extending outward 100–125 miles from the center with speeds of 75–100 mph. These weaker winds affected a much larger area and lasted for 10–15 hours at some locations along the upper Texas coast. Damage occurred over a wide area, but was not as severe as that caused by Hurricane Charley. On average, hurricane force winds (over 74 mph) extend outward 50–70 miles from the center of a storm.

Forward motion of the storm and how far inland the strong winds will penetrate also contribute to the amount of wind damage. Usually a major hurricane will lose 50 percent of its landfall wind speed within the first 12 hours of landfall. For example, a 120 mph hurricane typically weakens to 60 mph 12 hours after landfall. The faster the forward speed, the further inland the damaging winds will extend. In 1995, Hurricane Opal made landfall along the Florida panhandle moving northeast at 25–30 mph bringing hurricane force winds to Atlanta, Georgia.

Strong winds also contribute to storm surge, the gradual rise in the sea level prior to and during the landfall of a tropical cyclone (fig. 3.6). Storm surge height varies greatly from one location to another and from one storm to another. The factors affecting storm surge generation include, but are not limited to. offshore bathymetry, local coastal topography, storm intensity, angle and speed of approach, and the size or extent of hurricane force winds.

The offshore bathymetry and coastal topography vary from one coastal location to another. Along the Gulf Coast the offshore waters are fairly shallow allowing large storm surges to be generated. In contrast, the offshore shelf along the East Coast of the United States is much deeper counteracting the generation of large surges. Coastal topography greatly enhances surge levels given a favorable angle of approach. Higher storm surge tends to occur at the heads of inlets and concave coastlines where the storm surge becomes trapped. Regions particularly vulnerable to storm surge include Lake Pontchartrain and the Mississippi coast, Tampa Bay, Apalachicola Bay, the upper Texas and southwestern Louisiana coasts, the Chesapeake Bay, and portions of the lower Hudson River and Long Island.

The larger the wind field associated with a landfalling tropical cyclone the greater the storm surge will be. Larger storms are capable of moving larger quantities of water onto the coast. Large hurricanes such as Ike, Carla, Katrina, Rita, Ivan, and Opal have produced some of the most devastating storm surges along the Gulf Coast even though they have not

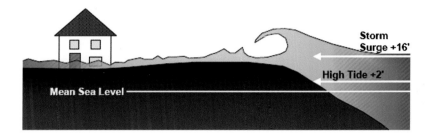

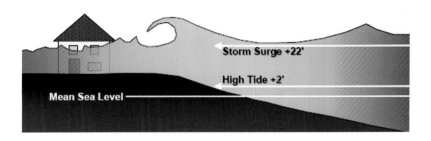

been the strongest hurricanes making landfall. While intensity plays a role in storm surge generation, the extent of the wind field is a much greater determining factor in surge extent and inundation.

Storm surge and Saffir-Simpson categories do not correlate with one another. For example, there is no such thing as a Category 2 storm surge or a Category 4 storm surge. Hurricanes Dolly and Ike were both Category 2 storms, but produced storm surges that were significantly different mainly because Ike was nearly four times larger than Dolly. Storm surge, compounded with wave action, can be extremely damaging and causes the majority of the coastal damage. The force of the wave action into structures along

Figure 3.6. The shallow bathymetry in the Gulf of Mexico contributes to heightened storm surge. Storm surge intensity is also increased if a hurricane makes landfall at high tide. Courtesy Rice University Archives.

with debris floating on top of the storm surge can result in significant damage to structures near the coast.

Tornadoes often accompany hurricanes making landfall, however they are typically weak and short-lived and when compared to the other damaging aspects of wind, storm surge, and inland flooding usually do not claim many lives. During landfall, tornadoes are typically generated in the right or eastern side with the most favorable location in the right front quadrant of a tropical cyclone. Within this area, the low level shear is maximized resulting in favorable low level rotation and tornado genesis. Tornadoes are most common in the feeder bands surrounding the eastern side of a tropical cyclone and not within the main rain shield. The stronger a storm is at landfall the greater the tornado potential.

For decades Hurricane Beulah held the record for producing 115 tornadoes across south Texas, but in 2004 Hurricane Ivan produced 127 tornadoes over a 2-day period across the southeastern United States. Tornadoes can affect locations up to several hundred miles from the point of landfall and extend well inland to locations well outside the impact zone. Most tornadoes in tropical cyclones are weak, typically ranging from EF0-EF1 on the Enhanced Fujita Tornado Damage Scale.

## Measuring and Predicting Rainfall

The single greatest factor when determining the rainfall amount from a tropical system is the forward speed of the system. A rule of thumb used to estimate the amount of rainfall a tropical cyclone will produce is to divide 100 by the speed of the forward motion of the hurricane. For example, a tropical system moving at 10 mph could be expected to produce upwards of 10 inches of rainfall. The resultant inland flooding can extend hundreds or even thousands of miles inland and last for several days after landfall as the tropical cyclone decays. The intensity of the system is usually not as important, although weaker systems tend to be more disorganized and move slower when they are decoupled from the main steering flow.

Flooding from tropical cyclones usually results from three factors: the rain shield of the system, feeder bands on the fringes, and nocturnal core rains. The large rain shield of a tropical system can cover hundreds of miles and last for several hours gradually saturating the ground and eventually leading to flooding. Feeder bands that develop along the outer periphery of the inner core rain shield can result in short-term, excessive rainfall rates and flash flooding. Feeder bands are notorious for exhibiting training cells that remain over the same locations for several hours. Such bands are most common on the right side of the tropical system and can extend several hundred miles from the center, resulting in flash flooding well away from the point of landfall.

The third phenomena is known as a nocturnal core rainfall event and is most common with decaying tropical systems, or mid level to low level low pressure systems within a tropical air mass. During the daylight hours, solar insulation breaks up the inflow into tropical systems decreasing convergence and resulting in the development of outer banding convection. In the evening, as the low level flow increases toward the center, convergence increases resulting in the development of thunderstorms around the center. These storms usually grow into a cluster and move slowly along near the center of circulation. Given strong inflow and copious amounts of moisture, extremely heavy rainfall totals can occur from these nighttime events leading to rapid flooding. Such events can occur well inland away from the landfall impact area and due to their nighttime initiation, make public warning difficult. There are several examples of this phenomena including Tropical Storm Charley (1998) near Del Rio, Texas.

## Conclusion

Tropical cyclones can impact large areas, causing rainfall and inland flooding, damaging winds, storm surge and wave action, and tornadoes. The Houston/Galveston Area is particularly prone to these types of storms. Recently, hurricanes Ike, Katrina, and Rita stand out as major producers of storm surge in the area. In addition, tropical storms, such as Allison (2001) and Erin (2007), can cause significant inland flooding. Chapter 4 discusses flood prediction and warning systems that have been developed in response to severe urban flooding.

# 4

# Flood Prediction and Flood Warning Systems

*Jeffrey Lindner, Dave C. Schwertz, Philip B. Bedient, and Nick Fang*

## Introduction

Floods and flash floods are among the leading causes of weather related deaths in the United States, resulting in 136 deaths per year and over $4.0 billion in property damage. With heavy rains and the continual threat of severe storms, the Gulf Coast region is particularly susceptible to flooding. Far from a declining hazard, population growth has caused expansion of residential and commercial areas deep within floodplains yielding ever greater property loss and more frequent damages (fig. 4.1). Urbanization alters the hydrological setting of watersheds resulting in faster watershed response times and higher peak flows.

Flood and flash flood forecasting skills have seen much less improvement when compared to tornado or hurricane forecasting. However, significant efforts were made to improve forecasting skills after the devastating flooding following Hurricane Floyd (1999) and the recent onslaught of inland flood damage from tropical cyclones making landfall on the Gulf Coast. Urban flash flood forecasting is the most difficult of all flood forecasting efforts due to the rapid response of urban watersheds and limited computer modeling ability, leading to untimely data when compared to flood forecasting for major river systems. Customized flood alert systems using radar, however, have seen some advance in Texas (Bedient et al. 2008).

## Hydrological Ingredients for a Flash Flood Event

The Gulf Coast endures frequent flooding events tied to many meteorological scenarios such as the landfall of tropical cyclones, the passage of frontal systems, and mesoscale phenomena, such as the sea-breeze front and outflow boundaries. Just as the moisture available in the atmosphere that produces heavy rainfall is important in determining a flood event, so too is what happens once the rainfall contacts the ground. Rainfall rates and duration are

Figure 4.1. Floodwaters caused by Tropical Storm Allison inundate trucks and cars trapped on a freeway. Tropical Storm Allison dumped over 24 inches of rain on Houston in 9 hours. Copyright © 2001 Dan Wallach. Used with permission.

key ingredients for a flash flood especially in an urban setting, but the type of ground the rain falls on, the time of year, and the moisture content of the soil all play an important role in determining how hydrological basins will respond to the rainfall.

The types of watersheds that exist along the Gulf Coast are varied and the effects of excessive rain may be felt far downstream. Generally, the elevation rise along the Gulf Coast is considered flat with some exception over south-central Texas where the terrain becomes slightly hillier. The flat terrain along the Gulf of Mexico causes

slow run-off and large areas of ponding, which usually yield a slow rise on bayous and creeks that are low-lying and interface with the marshlands. In the hills of Central Texas more rapid run-off occurs because soils are rockier, leading to flash flooding in normally dry creek beds. The corridor along I-35 from Del Rio northward through northern Texas is known as "flash flood alley" and is the site of multiple high water rescues and flooding deaths each year (figs. 4.2 and 4.3a-e.)

Urbanization around the large city centers along the Gulf Coast yields

**Flash Flood Alley**

The Central Texas Hill Country is the most flash flood-prone area of North America.

Dallas

Waco

Temple

Austin

San Marcos
New Braunfels
San Antonio

Copyright 2005
FloodSafety.com

Figure 4.2. The Central Texas Hill Country denoted in the white band is the most flash flood prone area in the United States. Rains originating in the hills (red area) quickly flow down gradient into major cities. Courtesy Marshall Frech. From the movie *Flash Flood Alley*, produced by Marshall Frech.

another interesting hydrological setting. Most of the larger urban centers are nestled in the flat coastal plains where drainage is already slow, due to a lack of elevation rise when compared to that of Central Texas. In addition, urban sprawl consisting of concrete freeways, supermarkets, and housing developments results in miles of impermeable ground cover. Thus, where storm water was historically allowed to collect and pond in marshes, there are now subdivisions.

When excessive rainfall occurs in an urban setting, the first lines of defense are the street gutters and roadside ditches that collect roadway storm water and discharge into area streams and watersheds. These systems have varying design capacities and are frequently overwhelmed during the warm-season months when short-term, excessive rainfall from daily sea-breeze can rapidly produce 2–3 inches of rain in less than an hour. Once the primary drainage system is at full capacity, storm water begins to collect and pond in low-lying areas.

Figure 4.3 a–e. While camping near the Frio River in the Texas Hill Country, a flash flood occurred at this camp site. The high water mark shown on the trees is estimated at 20 feet. Within hours the water levels receded to normal levels. Photos courtesy Tony, Gatesville Texas.

If excessive rainfall continues, the storm water will flow across the natural slope of the ground toward the lowest point of elevation, usually a stream or bayou.

The phenomena of overland flow and ponding are common along the Gulf Coast due to the combination of urbanization, relatively flat terrain, and frequently high short-term rainfall rates. The following sections discuss methods of predicting flooding.

## Flash Flood Forecasting by the National Weather Service

The National Weather Service (NWS) flash flood program is the responsibility of the local NWS field offices (fig. 4.4). With Texas leading the nation in flash flood deaths 21 out of the past 36 years, this is a program the NWS takes very seriously. They use a variety of products to alert the public about a flash flood threat. When conditions develop that may bring about flash flooding, the local NWS will issue a Flash Flood Watch. This is the first step in the watch/warning process and is usually done 12–24 hours prior to the onset of heavy rainfall. When flash flooding is imminent or observed, a flash flood warning is issued. The duration of these warnings usually ranges from 2–6 hours. Flash Flood Statements are issued as a fol-

low-up to the warning and provide updates of flood locations, rainfall estimates, and radar trends. The warnings are cancelled when the threat of flooding has ended. Another product that is used to warn of flooding is the Urban and Small Stream Flood Advisory. This advisory system alerts the public to the flooding of small creeks, streams, streets, and other low-lying places in urban areas where the rise in water is an inconvenience, but not life-threatening.

The Weather Forecast Offices (WFO) have two different software programs available to help analyze and assess the flash flood threat, the first of which is called Flash Flood Monitoring and Prediction (FFMP). This software based on a Geographic Information System (GIS) and makes use of customized drainage basins and gridded GIS-based flash flood guidance. The basin customization process can reduce the basin size down to as small as a few square miles and can add flow direction of the water from basin to basin. FFMP integrates gridded flash flood guidance, produced at the River Forecast Centers (RFC)and the Digital Hybrid Reflectivity (DHR) product, which is generated with every volume scan of the WSR-88D Doppler radar to give detailed information of where the heaviest rain has fallen and the given flash flood potential (fig. 4.5).

The WFO meteorologists evalu-

ate the FFMP output information received from the affected counties and radar performance data to decide on what advisory to issue. Once a flash flood warning or flood advisory is issued, frequent follow-up calls are made to the affected counties to obtain updates on the flooding and for verification purposes. Some municipalities have an automated gauge network that monitors low-water crossings and streams that frequently flood during heavy events. These networks provide valuable data to the local NWS Office and provide real-time feedback of the flood event.

The Site Specific Hydrologic Predictor Model (SSHPM) is the second software package available to the local NWS office to assist in the flash flood prediction program. This software is used by meteorologists to produce flood forecasts for the fast reacting streams and creeks that can go from base flow to high water in a matter of a few hours. The software incorporates

Figure 4.4. This facility houses the Houston/Galveston National Weather Service Office and the Galveston County Office of Emergency Management. It was specially constructed by the National Weather Service and Galveston County to withstand a Category 5 hurricane. Photo courtesy Nick Saum (www.nicksaum-photography.com).

Figure 4.5. Next-Generation Radar (NEXRAD), also known as the Weather Surveillance Radar, 1988 or Doppler Radar (WSR-88D), is operated by the National Weather Service and National Oceanic and Atmospheric Administration (NOAA), enabling accurate weather forecasting. Photo courtesy NOAA.

a rainfall-runoff model calibrated for the drainage basin, observed and forecast rainfall, and real-time gauge data (if available). The gauge and observed rainfall data are automatically read by the software, while meteorologists input forecast rainfall data. As adjustments are made to the observed and forecast rainfall, the forecast hydrograph is instantly updated to reflect those changes.

When an acceptable forecast is reached, the data is saved and Flood Warning (FLW) product is issued using the Riverpro software. Since these forecasts are for creeks and streams, flood stages and moderate and major categories are set for the forecast point.

Follow-up forecasts and cancellation messages are issued using the Flood Statement Product (FLS). If the creek or stream is forecast to have a significant rise but not reach flood stage, a Hydrologic Statement (RVS) can be issued to cover the event.

While large watersheds respond more slowly to excessive rainfall, urban watersheds can respond very quickly, reducing the time allowed for RFC and NWS forecasters to run complex computer models to determine flood flows and to develop a crest forecast. In many urban environments the watersheds are small and the run-off instantaneous, yielding a difficult warning process and short warning lead times. In many cases, the flood warning for a particular watershed in an urban environment may be reactive instead of proactive due to a lack of timely data and the rapid response of the watershed.

**Flash Flood Forecasting in the Urban Setting**

The development, expansion, and increasing data quality of stream gauging systems such as US Geological Survey (USGS) monitoring sites and flood alert systems, along with better rainfall reports and hourly estimated precipitation amounts from the RFC, have allowed substantial increases in the flood warning decision arena. The use of multiple data sources and interagency coordination among local NWS offices, emergency managers, law enforcement, and local water management districts can expedite the flow of critical information to the public, reducing property damage and saving lives during rapidly evolving flooding situations in urban environments.

While public warnings may be issued in a timely manner prior to the onset of flooding conditions, one must account for the fact that the majority of the public respond differently to flash flood events than to other severe weather products. Citizens must not only first receive the warning product, but must also understand the terminology within the warning and have an accurately perceived level of risk. The public must realize the dangers of floodwaters and the amount of force only a small amount of flowing water can exhibit. Even when flash flooding and heavy rainfall are in progress, the public takes little action to protect themselves from the dangers that exist and continue to drive into swollen low-water crossings and flooded roadways (fig. 4.6).

Flash flood and flood watches and warnings are broadcast through the same media as tornado and hurricane watches and warnings, such as NOAA Weather Radio, public media stations, local radio stations, and the

Figure 4.6. A Houston freeway holds rising floodwaters, as a trailer turns at right angles to its tractor. The water level rose quickly, lifting the trailer. Moving water only 18–24 inches deep will cause a car to float. Copyright © 2001 Dan Wallach. Used with permission.

Emergency Managers Weather Information Network (EMWIN) paging system. However, the majority of citizens do not realize the risk or severity of a flash flood until they either experience the event first hand or are able to view the inundation through live television broadcast. While region-wide watches and warnings spanning multiple counties are generally understood, more specific warning information for a particular watershed reach is not.

Residents are largely unaware of which watershed they are located within, especially in regions with multiple small watersheds in which they may live, work, and commute across multiple watersheds. Warnings for such instances may span the entire watershed or just a small section and tend to have short lead times due to the fast response of smaller watersheds. Cities such as Houston, Austin, and San Antonio are comprised of several small creeks and bayous that, during times of excessive rainfall, can each have their own flooding problems requiring specific warning information. Inclusion of familiar landmarks such as major street intersections, shopping malls, and historical sites help citizens utilize the warning product better. The addition of historical crest data also provides information to residents along a certain watershed that can relate to the historical flooding of the past.

The development and launch of the Advance Hydrological Prediction Service (AHPS) has brought a mapping and graphical approach to flood warnings which provides invaluable data to residents via the internet. Products include the current stage and forecast hydrograph at a forecast point, categorical flood levels including minor, moderate, major, and record flood stage, along with definitions and impact statements at each level. The recent addition of inundation maps for several forecast point locations along the Gulf Coast yields detailed floodplain maps as well as inundation levels at various stages. While the technology available to accurately forecast flood hazards continues to improve, the dialogue between the forecasters, emergency management, and the public remains a crucial component in avoiding flood-related disasters.

## Customized Flood Alert Systems in the Texas Medical Center

In urban watersheds throughout the United States, web-based, customized flood alert/warning systems are becoming more prevalent, and are can provide the information necessary to mobilize the evacuation of people and property prior to flooding in critical areas. These systems are designed to collect, process, analyze, and disseminate hydrologic information, ideally in real time, for the purpose of providing accurate advance warnings of an impending flood condition. The growth of these systems is due to the increased availability of high-quality hydrologic data and web delivery systems. These systems are very useful along the Gulf Coast for flood warnings (Bedient et al. 2008), permitting radar-based flood alert systems to successfully issue flood warnings at specific locations with increasing lead-time at even the small (100 square mile) urban watershed scale. Once installed and accepted as part of a flood management system, time-critical actions, which may result in significant costs to both public and private organizations, can be taken with ample time to minimize flood damages.

The original FAS, developed in 1997–98, converted radar rainfall into watershed rainfall. These real-time rainfall totals were entered into an active nomograph for flood peak prediction. The system was used by the Texas Medical Center (TMC) in the Brays Bayou watershed to disseminate flood warning information to its 22 member institutions and hospitals, to develop plans for emergency response to severe weather events, and to implement these flood responses. The FAS proved successful in several small storm events from 1998–2003 and two major storm events in 1998 (Tropical Storm Frances) and 2001 (Tropical Storm Allison).

The second generation of the Rice/Texas Medical Center Flood Alert System, known as FAS2, was developed with the aim of addressing many of the shortfalls of the original system. FAS2 utilizes 5-minute NEXRAD radar rainfall data (Level II) coupled with two real-time hydrologic models, a lumped model (RTHEC-1) and a distributed model (Vflo™). These models generate flow hydrographs as storms progress and deliver warnings in a readily understood format to TMC facility personnel with 2–3 hours of lead time (fig. 4.7). Improving the accuracy of the system was accomplished in a variety of ways, including improved radar data input and resolution, real-time calibration of the radar with rain gauges when available, and the development of real-time hydrologic models. These models take greater advantage of the spatial and temporal distribution of real-time radar data to provide significant lead-time improvements by estimating when peak flows would actually occur. Information dissemination tools in addition to the internet were added to the new system, including automated notification via cellular phones, pagers, and email.

The system generally predicts floods with average differences of 0.87 hour in peak timing and 2.8 percent in peak flows. Particularly, FAS2 was found to better predict shapes, peak values, and peak timing for bigger storm events

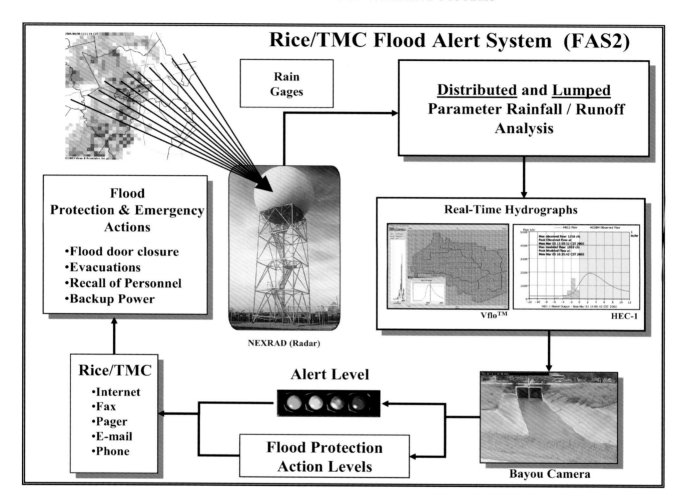

Figure 4.7. The Rice/TMC Flood Alert System (FAS2) uses data collected from NEXRAD radar, rain gauges, and elevation maps to create hydrographs that show the progress of a storm in real time. At pre-determined levels, system users are alerted to begin taking flood protection measures. Courtesy Rice University Archives.

than for small and/or moderate events. For instance, it accurately predicted the peak flow rate at 26,811 cubic feet per second when the observed flow rate was 25,500 cubic feet per second during Hurricane Ike on September 13, 2008, a 5 percent peak difference (fig. 4.8). Overall, the previous performance of FAS2 has proven that it is a reliable and operational flood warning tool, providing confidence to emergency personnel in the Texas Medical Center and Rice University during any event.

Due to its excellent performance in the past, the system has been upgraded to a more advanced platform (FAS3), which provides better calibrated

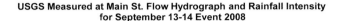

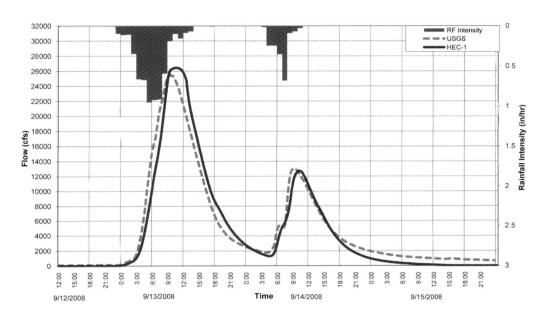

Figure 4.8. During Hurricane Ike, the predicted and the observed hydrographs were very similar. The accuracy of the FAS2 system allowed the Medical Center to stay open during Hurricane Ike, preventing significant losses. Courtesy Rice University Archives.

hydrologic models and incorporates a new hydraulic prediction tool: Flood-Plain Map Library (FPML) in Google Maps. The FPML module aims to provide visualized information using over 60 floodplain maps in Google Maps instead of just showing hydrographs in real time based on radar rainfall data. The Texas Department of Transportation (TxDOT) implemented such a system for a critical transportation corridor along a major evacuation route in Houston, Texas. Similar flood warning systems are also being used as a prototype for other flood-prone areas such as the City of Sugar Land and Austin, Texas.

## Conclusions

Flood warning and alert systems are vital for the protection of flood prone communities. Their success is highly dependent upon accurate predictions and comprehensive communication systems. In extending flood alert systems to coastal areas, it will be necessary to incorporate the effects of storm surge on inland flooding. Hurricane Ike clearly demonstrated this need for Bolivar Peninsula and Galveston Island, Clear Lake, and the Houston Ship Channel area in September 2008. The following chapter discusses the advances in storm surge prediction and modeling.

# 5

# Predicting Storm Surge

*Clint Dawson and Jennifer Proft*

## Introduction

Figure 5.1. As the storm surge from Hurricane Ike continues to rise, Bill Murphy, second from right, waits with three rescuers for a boat to pull them to safety after Murphy's wife Barbara and two others were rescued by a Coast Guard helicopter in High Island, Texas. Photo courtesy Guy Reynolds/Dallas Morning News.

Every hurricane has the potential to inflict damage in one or all of the following ways: wind, rainfall, tornadoes, and surge. Of these, storm surge has been responsible for some of the most devastating hurricane-related damage. Storm surge occurs when sea levels rise in the face of low barometric pressure. The resulting mass of water is pushed onto shore by strong hurricane winds as described in chapter 3 (fig. 5.1). Long known for its damaging effects, storm surge is difficult to predict and has been responsible for the loss of thousands of lives and billions of dollars in damages along the Texas and Louisiana Gulf coasts. Sea walls, levees, bulkheads and the like have all been built in attempts to protect lives and communities from the onslaught of this powerful force. However, as residents move towards coastal areas in greater numbers, the potential for significant loss continues to increase.

As in any natural hazard scenario, the safety of a community is directly

tied to the ability of forecasters to successfully predict the location and timing of storm surge and respond. To do this, it is necessary to understand the causes and physical effects of surge, both on the meso- and the micro-scale. Recently, computer modeling has become an effective tool for studying storm surge mitigation. Modeling lets forecasters predict the location and severity of storm surge prior to hurricane landfall. Prior to the advent of computer-based modeling, one could only make an educated guess based upon historic, empirically observed data (Resio et al. 2008). For example, if a historical Category 2 hurricane resulted in a 5-foot storm surge, a future storm with similar wind speeds would be expected to have roughly the same surge effects.

The inaccuracies associated with such observational analyses are evident given the myriad of complex factors involved in the formation of hurricane storm surge. These factors include wind speed, wave-current interaction, tides, atmospheric pressure, riverine flows, rainfall, wetting and drying capacities, and local topography and bathymetry (Bunya et al. 2010). They are best understood through computational methods. Given the complexity of such factors, attempts to understand hurricane storm surge by observation alone invariably fall short. The benefits of storm surge modeling are easily

characterized and fall into one or more of the following categories: structural mitigation, land use, long-term land development, and evacuation.

**The Benefits of Storm Surge Modeling**

Sea walls, dikes, levees, flood gates and the like have historically played a role in protecting vulnerable communities from the effects of storm surge. However, simply placing a solid structure of an arbitrary height between a community and the sea is a dangerous way to go about protecting both people and property (fig. 5.2). During Hurricane Katrina, the levees in New Orleans were not only breached, but the Arpent Canal Levee failed completely and approximately 131 billion gallons of water flooded the city. Computer-based simulations allow forecasters to model various storm surge heights against proposed structural mitigation, testing their response and enabling city officials and emergency management to make educated decisions about what type of structure, if any, will protect communities in a particular area.

While structural mitigation plays an important role in storm surge management, equally important is the role of land use and non-structural mitigation. Where and what we choose to build may have far reaching effects,

Figure 5.2. Piles of debris are lined up along the seawall on Galveston Island where Hurricane Ike made landfall. Photo courtesy Robert Kaufmann/FEMA.

either increasing or decreasing the damage storm surge can inflict. For example, coastal wetlands are able to attenuate the large volumes of water from storm surge, resulting in reduced surge elevations further inland. Ecological preservation and wetland restoration in coastal areas is one way to mitigate the effect of storm surge. These areas have an incredible capacity to slow the forward motion of surge, as well as store the water after the event. Computer-based modeling allows planners and policy makers to better understand which areas function well as natural barriers to the sea and limit development in those areas.

Land use decisions in areas vulnerable to storm surge are an important step in protecting communities from severe damage. However, it is difficult to describe spatially which areas are at greater risk to storm surge damage than others. Land contours, elevation, proximity to the coast, and prior development all play a role in determining which areas are more prone to severe storm surge. It is the task of storm surge modelers to predict which areas will fare best and forecast the storm surge severity, given a unique hurricane scenario and location. Planners use the resulting information to make informed decisions as to the size, type, and location of development in coastal areas.

Those living in coastal communities will always be exposed to the potential for damaging surge during a hurricane or severe storm event. It is imperative to understand where the storm surge from a particular storm will have the greatest effect in order to help protect coastal communities and prevent deaths. Storm surge modeling allows forecasters to predict a hurricane in real time, helping decision makers to efficiently evacuate at-risk populations with ample lead time ahead of landfall. Social vulnerabilities and evacuation are further discussed in chapters 6–8.

Accurate storm surge modeling is invaluable in the planning processes both far out and near the time of a severe storm event. The SSPEED Center is currently using the Advanced Circulation (ADCIRC) model and other methods to evaluate community protection through structural and non-structural mitigation. In the following sections, the governing processes of modeling storm surge are explained, specifically in the use of the ADCIRC model.

## The Evolution of the ADCIRC Model

As previously mentioned, computational methods necessary to analyze storm surge are a relatively recent introduction. Prior to the development of advanced technology, empirical data gathered by eyewitnesses were the principle means for assessing local risk to storm surge. This method is problematic because severe storm surge occurs relatively infrequently in any particular area making it very difficult to predict a future event given that it may be very different from its predecessor. Due to these empirical limitations, the mechanics of storm surge development were not well understood before the 1960s. The sudden availability of digital computation provided a way to input quantitative storm surge equations. By running computer simulations, scientists were able to predict hurricane impacts and storm surge over large areas (Resio et al. 2008).

The National Hurricane Center (NHC) and the National Oceanic and Atmospheric Assocaition (NOAA) collaborated to develop the Sea, Land and Overland Surges from Hurricanes (SLOSH) model. It allows forecasters to estimate the potential maximum surge depth for an area to approximately 20 percent accuracy. The SLOSH model incorporates data on the pressure, size, forward speed, track, and wind speed of a hurricane. While such early simulations allowed for a greater understanding of storm surge as a hurricane made landfall, behavior at a local level was still difficult to predict. This improved when, in the early 1970s, space-based remote sensing technology became available. The technology known as Light Detection and Ranging (LIDAR) allows the creation of multi-cell topographic grids that express both local geography and bathymetry (fig. 5.3). These are applied to storm surge models as quantitative inputs to make even more accurate predictions. However, the accuracy of the SLOSH model is dependent upon the accuracy of the predicted hurricane track. If the hurricane does not follow the forecasted path, the SLOSH model may be inaccurate and the landfall data irrelevant for local authorities and decision makers.

The ADCIRC storm surge model was developed as a tool to study the effects of a potential catastrophic storm on southern Louisiana (Bunya et al. 2010). As a result of this effort, a high resolution grid modeling the southern Louisiana coast was developed prior to Hurricane Katrina. After hurricanes Katrina and Rita, the model was expanded to include high resolution elevations for the Louisiana, Mississippi and Alabama coasts (Dietrich et al. 2010), and it was used to successfully hindcast both storms. Recently, a high resolution grid of Texas has been developed for the purpose of using the ADCIRC model to update federal digital flood insurance rate maps along the Texas coast and in recent studies in the wake of Hurricane Ike (Kennedy et al. 2010) (fig. 5.4). Furthermore, an ADCIRC forecast model, the ADCIRC Surge Guidance System (ASGS; Fleming et al. 2009), has been developed as a predictive tool as storms approach land. This model may be used by emergency managers in conjunction with SLOSH.

Today, ADCIRC is primarily used by the Army Corps of Engineers, The United States Naval Research Laboratory, the National Oceanic and Atmospheric Administration, the National Civil Engineering Laboratory, and the Federal Emergency Management Agency. The model is continually being developed and updated by research groups at The University of North Carolina at Chapel Hill, The University of Notre Dame, The University of Oklahoma at Norman, and the University of Texas at Austin (http://www.adcirc.org).

Figure 5.3. LIDAR data is collected by a sensor that emits a high frequency laser beam through an opening in the bottom of an aircraft. The sensor records the time difference between the emission and the return of the laser signal to determine the elevation of the land below. Courtesy Rice University Archives.

Figure 5.4. The ADCIRC Model Texas Grid.

## Modeling with ADCIRC

One of the major difficulties in changing popular opinion of storm surge severity arises from the misperception that storm surge height is directly related to average hurricane wind speed. Storm surge height can vary greatly within a single category on the Saffir-Simpson Wind Scale. The potential danger of this misperception is appropriately expressed in an anecdote arguing that perceptions of Hurricane Camille caused more deaths in 2005 than the actual storm in 1969 (Resio and Westerink 2008).

Hurricane Camille made landfall in Mississippi as a Category 5 in August 1969. In 2005, residents who had been safe from storm surge during Hurricane Camille assumed that they would be just as secure, if not more so, during Hurricane Katrina, a Category 3 storm. However, Hurricane Katrina caused much more severe storm surge, affecting a greater area and surprising residents who had assumed surge severity was directly reflected in hurricane category.

The ADCIRC model can predict storm surge height and affected area before a hurricane makes landfall indicating which communities are at risk. It calls for a multitude of data input, each factor contributing to increased prediction accuracy. The following primary forcing functions and parameters are included in the model and are discussed in the subsequent paragraphs: wind speed, atmospheric pressure, wave current interaction, tides, wetting and drying capabilities, riverine flows and rainfall, and local topography and bathymetry.

The faster the wind speed of a hurricane, the higher the wind forcing. Wind forcing causes the water in front of the storm, typically between the storm and shoreline, to pile upon itself, increasing the water depth preceding the storm. Wind speed measurements are obtained both through direct observation and an analysis that combines inner and outer core wind speeds to create wind field values. These values are then assigned to specific locations on a wind speed grid. At all other locations on the grid, the wind field is interpolated from known wind speeds.

Atmospheric pressure also has an effect on the mass of water piled up and pushed towards shore in front of a storm. In order to maintain a constant pressure below the sea surface at a particular point, surface water exposed to the low pressure of a hurricane will rise, increasing the depth of water. The effect is estimated at a 10 millimeter (0.39 inch) increase in sea level for every 1 millibar decrease in atmospheric pressure (Harris 1963). Within the Gulf of Mexico, there are several different currents. The proximity of a hurricane to these currents has a direct effect on the severity of the storm surge. The energy of a breaking wave is exponentially related to the speed of the current in which the wave is traveling.

In addition to currents, the tidal dynamics at the time of impact play a significant role in the severity of the storm surge. The height of storm surge is measured above the height of the normal predicted tides. If a hurricane makes landfall at high tide, the height of water experienced on land will be significantly higher than at low tide. Residents around Galveston Bay experienced this phenomenon when Hurricane Ike made landfall at high tide, exacerbating the effects of storm surge (fig. 5.5). The large-scale behavior of oceanic water is affected by the interaction of tides on shore. This results in land areas alternating between being wet and dry. In order to model storm surge in a more physically realistic simulation, these characteris-

Figure 5.5. Winds from Hurricane Ike bent stop signs, flooded houses, and beached boats throughout this Galveston Island community. The Federal Emergency Management Agency and federal and state partners are working to assist businesses, individuals, and local governments after Hurricane Ike. Photo courtesy Leif Skoogfors/FEMA.

tics are also included in the ADCIRC model. Furthermore, increased riverine flows into estuaries may result from increased rainfall on land during a hurricane event. This increased flow can substantially increase water depth near local sea outlets such as Galveston Bay, also increasing storm surge severity.

Perhaps one of the most significant factors contributing to localized storm surge effects is the topography of the coastal land and the bathymetry of the sea floor near a storm surge event. Storm surge is greatly affected by the depth and shape of the sea floor and the type of land cover it encounters as it breaks onto shore. For example,

the height of storm surge in the Gulf of Mexico is much higher than in East Florida because the waters are much shallower.

The refinement of numeric elevation values has progressed along with the ability to accurately map these areas, particularly since the advent of LIDAR has enabled the creation of finite element grids in which each node possesses a value indicative of topography. The use of finite element grids allows for higher resolution in regions sensitive to storm surge, while allowing for less resolved regions out towards the open ocean. The grid used in the ADCIRC model for the Texas Gulf Coast contains over 6 million nodes with resolutions as high as 20 square meters near the shoreline, allowing modelers to take topographical micro-scale characteristics into account, yielding accurate local results. As state-of-the-art storm surge modeling continues to advance, our skill to perform highly accurate computational simulations increases. The most recent modeling successes were witnessed during hurricanes Katrina, Rita, Ike, and Alex.

**Forecasting Hurricane Ike**

Hurricane Ike gave modelers the opportunity to test their abilities. They were able to forecast storm surge before, during, and after landfall and were later able to compare those results to observed data at various points as a means of gauging their accuracy. A visual chronology of storm surge predictions for Hurricane Ike is presented as captured by ADCIRC (fig. 5.6 a-d).

**Modeling Alternative Scenarios**

The uses of storm surge modeling are not limited to isolated events; on the contrary, such modeling enables community-based mitigation decisions. Hurricane Ike made landfall at Galveston Bay as a Category 2 storm, but brought larger than expected storm surge, raising the question: "How can we protect the Texas Gulf Coast in the event of 'The Big One'?" Government officials and citizens wondered what would happen if a larger, more powerful hurricane was to make landfall at the same location in the future. After validating the ADCIRC model on Hurricane Ike, it is not only possible to simulate larger storms in the same region, but also the benefits of different proposed mitigation structures along the Texas coast.

In the wake of Hurricane Ike, AD-CIRC modelers at the University of Texas chose seven hypothetical landfall locations and one original location identified in fig. 5.7a to assess the point at which Hurricane Ike would have produced the greatest storm surge. It was found that the worst case scenario would have occurred had Ike made landfall at Point 7. The original path of Hurricane Ike, along with worst case scenario, is illustrated in fig. 5.7b. This allowed the ADCIRC modelers to accurately compare storm surge at multiple observation points for different storm scenarios. The ADCIRC results for Hurricane Ike modeled at the original landfall and at Point 7 are shown in figs. 5.8 a and b.

The "Ike Dike" was proposed by Dr. William Merrell (2010) as a permanent solution to the damaging effects of storm surge on Galveston Bay and the Houston Ship Channel. It would span the length from the west end of

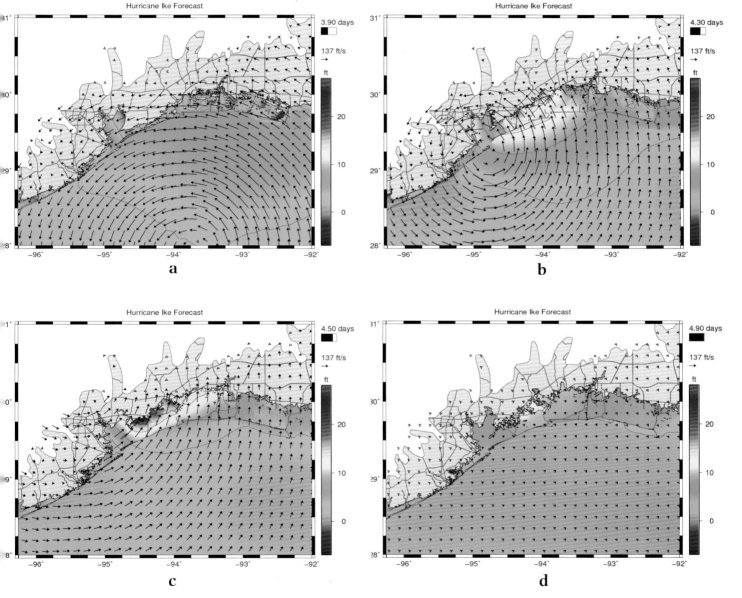

Figure 5.6a–d. The ADCIRC model maps the progression of Hurricane Ike as it makes landfall, showing wind vectors and inundation levels along the Texas coast.

Figure 5.7a. ADCIRC modelers at the University of Texas assessed eight locations at which Hurricane Ike could have made landfall. Had Ike made landfall at Point 7, effects would have been much more severe than what occurred. Courtesy Rice University Archives.

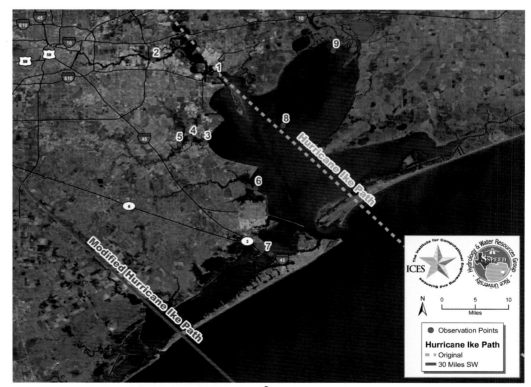

Figure 5.7b. Hurricane Ike's actual path mapped alongside the path that Ike could have taken from Point 7. The resulting inundations are given in table 5.1. Courtesy Rice University Archives.

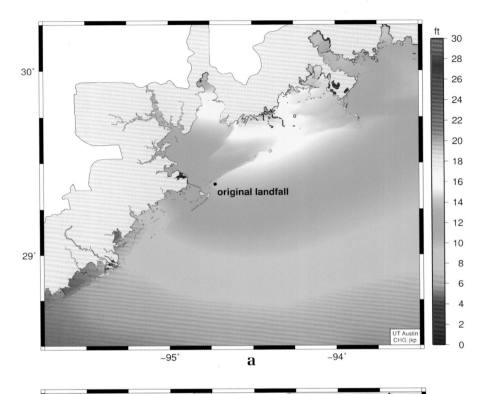

a

Figure 5.8a. The maximum inundation that would be caused by Hurricane Ike if it had made landfall at the original location without the proposed Ike Dike in place.

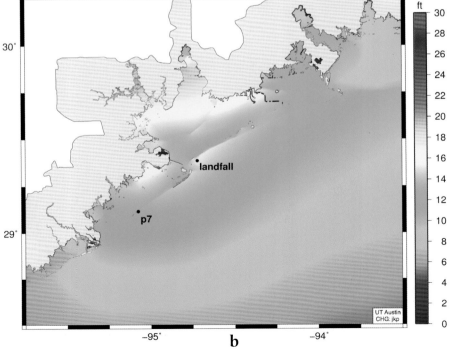

b

Figure 5.8b. The maximum inundation that would be caused by Hurricane Ike if it had made landfall at Point 7 without the proposed Ike Dike in place.

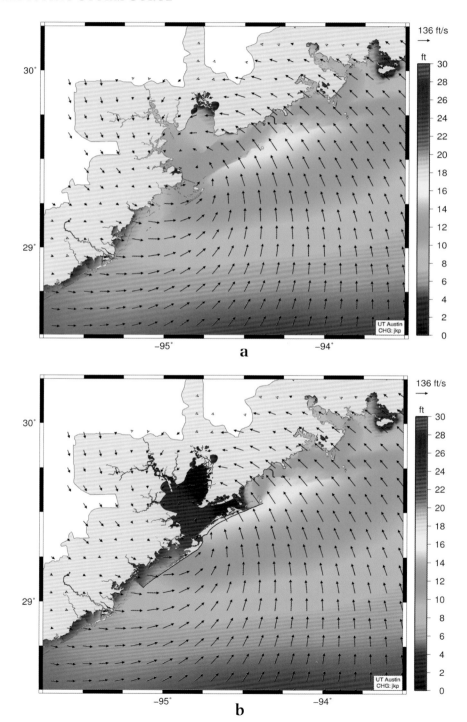

Figure 5.9a–b. Water elevation and velocity vectors as Hurricane Ike makes landfall along the Texas coast, with and without the Ike Dike in place.

Galveston Island all the way to the eastern end of the Bolivar Peninsula. The question that remains is whether the Ike Dike would have an effect against a more powerful storm surge. ADCIRC modelers computed scenarios with and without the Ike Dike in place (fig. 5.9a, b). Storm surge heights at various locations in the region for each path are given in tables 5.1 and 5.2. The results show that the Ike Dike could reduce storm surge impacts by as much as 10 feet. However, the Ike Dike stands among the most costly and time-consuming proposals.

As another structural alternative, the SSPEED Center has proposed constructing a flood gate across the Houston Ship Channel near the Hartman Bridge (fig. 5.10). This option may be much more affordable, while still providing economic and environmental sustainability to the region. ADCIRC modeling has allowed researchers to examine the comparative benefits of such structural proposals, while continuing to consider the economic costs. Such quantitative information is invaluable to decision-makers and serves as a visual illustration of possible solutions to a serious threat.

Despite the incredible capacity of ADCIRC modeling, it is important to remember that modeling alone cannot function as a decision-making tool. It is still necessary to determine whether cost of structural mitigation outstrips

**Table 5.1. Hurricane Ike storm surge modeled at original landfall and Point 7 without Ike Dike.**

| Point | Location | Original Landfall | | Point 7 Landfall | |
|-------|----------|-------------------|--|------------------|--|
| | | Max (ft) | Date/Time | Max (ft) | Date/Time |
| 1 | Houston Ship Channel Entrance | 13.0 | 9/13 12:00 | 17.8 | 9/13 12:00 |
| 2 | Port of Houston | 13.5 | 9/13 12:00 | 19.0 | 9/13 12:00 |
| 3 | Kemah | 11.5 | 9/13 12:00 | 15.1 | 9/13 09:00 |
| 4 | Clear Lake at Taylor Lake | 11.5 | 9/13 12:00 | 15.1 | 9/13 09:00 |
| 5 | League City | 11.5 | 9/13 12:00 | 15.2 | 9/13 09:00 |
| 6 | Texas City Levee | 11.5 | 9/13 06:00 | 14.0 | 9/13 06:00 |
| 7 | I-45 Bridge | 12.0 | 9/13 06:00 | 13.5 | 9/13 09:00 |
| 8 | Galveston Bay | 12.0 | 9/13 12:00 | 15.0 | 9/13 12:00 |
| 9 | Upper Trinity Bay | 15.5 | 9/13 12:00 | 18.0 | 9/13 12:00 |

**Table 5.2. Hurricane Ike storm surge modeled at original landfall and Point 7 with Ike Dike.**

| Point | Location | Original Landfall | | Point 7 Landfall | |
|---|---|---|---|---|---|
| | | Max (ft) | Date/Time | Max (ft) | Date/Time |
| 1 | Houston Ship Channel Entrance | 2.5 | 9/13 12:00 | 7.0 | 9/13 09:00 |
| 2 | Port of Houston | 2.0 | 9/13 12:00 | 9.5 | 9/13 12:00 |
| 3 | Kemah | 3.0 | 9/13 06:00 | 6.5 | 9/13 06:00 |
| 4 | Clear Lake at Taylor Lake | 2.5 | 9/13 09:00 | 6.0 | 9/13 09:00 |
| 5 | League City | 2.5 | 9/13 09:00 | 6.0 | 9/13 09:00 |
| 6 | Texas City Levee | 4.0 | 9/13 06:00 | 5.0 | 9/13 06:00 |
| 7 | I-45 Bridge | 4.5 | 9/13 06:00 | 3.5 | 9/13 12:00 |
| 8 | Galveston Bay | 1.0 | 9/13 06:00 | 2.5 | 9/13 06:00 |
| 9 | Upper Trinity Bay | 4.5 | 9/13 12:00 | 5.2 | 9/13 12:00 |

Figure 5.10. There is a proposal to protect the Houston Ship Channel from storm surge by building a Rotterdam-like gate across the mouth of the channel.

its economic and environmental value. To answer such questions modelers, planners, emergency managers, government officials, and community members must all work in concert to implement effective solutions.

## Conclusion

As with any type of predictive modeling, modelers continue to seek improvements to ADCIRC and other models. With advanced computer and software technology, the computation time continues to decrease allowing for more efficient decision-making as a hurricane approaches the coast. Computational methodology also continues to improve both in computational speed and accuracy allowing for more varied model runs. With such improvements, the effect of micro or macro changes in land cover and resulting effects on storm surge can be modeled. Currently, this is a very labor-intensive process, which may eventually be expedited with the implementation of software templates.

The University of Texas at Austin team led by C. Dawson and J. Proft ran the ADCIRC Surge Guidance System (ASGS) for Hurricane Alex. The system uses the ADCIRC model combined with hurricane data obtained from the NHC to compute storm surge "now casts" or forecasts with each new advisory. The model uses a high resolution grid of the Texas coast developed by the ADCIRC group. Computational resources at the Texas Advanced Computing Center allow the ASGS to compute new forecasts within 2 hours of the release of each advisory from the NHC. This represents the first attempt at high-resolution, real-time forecasts of hurricane storm surge for storms impacting Texas. The results of the simulations were transmitted to Dr. Gordon Wells at the State Operations Center and used in emergency response planning and deployment of first response teams.

Storm surge modeling is a valuable tool that forecasts the short-term effects of an approaching storm, as well as the long-term effects of structural mitigation proposals. Models, such as ADCIRC, are an invaluable tool for reducing the risk of coastal communities to storm surge by enabling policy makers to make educated decisions about development, as well as life-saving decisions about evacuation prior to the landfall of a storm.

# 6

# Using Social Vulnerability Mapping to Enhance Coastal Community Resiliency in Texas

*Walter Gillis Peacock, Shannon Van Zandt,*
*Dustin Henry, Himanshu Grover, and Wesley Highfield*

## Introduction

Disasters like Hurricane Ike, as well as severe storms such as Allison, Katrina, and Rita are often referred to as "natural" disasters. Rather than being wholly "natural," however, these disasters result from the interaction among biophysical systems, human systems, and their built environment. Indeed, the emerging scientific consensus states that the damage incurred, in both human and financial terms, is largely due to human action or, more often, inaction (Mileti 1999). Communities in the United States and much of the world continue to develop and expand into high hazard areas. This contributes to increased hazard exposure and often results in the destruction of environmental resources such as wetlands, often increasing losses. In other words, many of the communities in our nation are becoming ever more *vulnerable* to "natural" hazards while simultaneously becoming less disaster *resilient*.

When disaster strikes, its impact is not just a function of its magnitude and where it strikes. Galveston, like most communities, is not homogeneous, but rather contains areas characterized by wealth, leisure, and privilege, as well as neighborhoods plagued by poverty, crime, and unemployment. Development patterns typified by sprawl, concentrated poverty and segregation shape urban environments in ways that isolate vulnerable populations. Severe storms like Ike are not "equal opportunity" events. These events affect different groups in different ways. Very often, the social geography interacts with the physical geography to expose vulnerable populations to greater risk.

In its broadest sense, community

vulnerability describes the susceptibility of a community and its constituent parts to the harmful effects of disasters. Variation in existing vulnerabilities influences the exposure of households, businesses, and communities to effects of natural hazards as well as the capacity and resources available to respond to and recover from disasters. While some can easily anticipate and respond to hazard threats, others find it more difficult if not impossible. As a result, in the aftermath of a severe storm, recovery can be highly uneven with only some parts of a community recovering quickly. The uneven nature of recovery can jeopardize the overall vitality and resiliency of a community and bring its future into question.

Using Galveston as a living laboratory, we will describe the ability of coastal communities and their populations (individuals and households) to anticipate, respond to, resist, and recover from disasters like Hurricane Ike. To illustrate, we use a recently developed internet tool, the *Texas Coastal Community Planning Atlas* (coastal atlas.tamu.edu). We conclude that undertaking a spatial analysis of social vulnerability should be a critical element in emergency management, hazard mitigation, and disaster recovery planning, helping communities reduce losses, enhance response and recovery, and strengthen community resilience.

## Social Vulnerability (SV)

When considering natural hazards, *vulnerability* generally refers to susceptibility or potential for experiencing the harmful impacts of a hazard event (Cutter 1996; Mitchell 1989). The foundation of vulnerability analysis, a hazards assessment, generally focuses on a community's exposure to hazard agents such as floods, surge, wave action, or winds (Deyle et al. 1998; NRC 2006:72–3). Such assessments identify the potential exposure of populations, businesses, and the built environment (housing, infrastructure, critical facilities, and so on). Also important are the physical characteristics of the built environment such as the wind design features of buildings, the height of structures relative to potential floods, as well as natural and engineered environmental features such as wetlands, dams, levees, or sea walls. These characteristics modify vulnerabilities and concomitant risk. As disaster and hazard researchers critically examine the nature and distribution of disaster impacts, the factors shaping the variability in exposure and access to technology which can mitigate impacts (i.e., shutters, impact-resistant glazing etc.), it becomes clear that more than just hazard exposure and the built and natural environment are shaping vulnerability. A new perspective emerges suggesting that social structures and processes also shape vulnerability; hence, the term *social vulnerability* (SV).

Social vulnerability is defined by Blaikie and others (1994:9) as "the characteristics of a person or group in terms of their capacity to anticipate, cope with, resist, and recover from the impacts of a natural hazard." An SV perspective focuses attention on the characteristics and diversity of populations in terms of broader social, cultural, and economic factors that shape the ability to anticipate future events, respond to warnings, cope with, and recover from disaster impacts. As the SV literature continues to grow, it has examined a variety of hazard and disaster contexts, identifying dimensions of social vulnerability related to race/ethnicity (Bolin 1986; Bolin and Bolton 1986; Perry and Mushkatel 1986; Peacock et al. 1997; Bolin and Stanford 1998; Fothergill et al. 1999; Lindell and Perry 2004), income and poverty (Peacock et al. 1997; Dash et al. 1997; Fothergill and Peek 2004), gender (Enarson and Morrow 1997, 1998; Fothergill et al. 1999) as well as a host of other factors such as age, education, religion, social isolation, housing tenure, etc. Very often, these factors are present in combinations (both poor and black, for example), which can exacerbate vulnerability (Morrow 1999).

Often, policies and practices related to disaster response assume that all residents of an area have the same information, resources, and ability to act. Further, they assume that all residents will react in the same way. Vulnerability factors, however, can shape and influence access to and knowledge of resources (physical, financial, and social), control of these resources, as well as perceived or real power within the larger community or society. They may also influence the capacity of the individual or household to act. For example, African-Americans often rely on social connections rather than the media or government to obtain information about threats or hazards (Perry and Lindell 1991; Morrow 1997). Even if a resident has the same information, he or she may not have the capacity (a car, for example) to evacuate in a timely manner. Renters are typically more mobile or transient than home owners and may not have local family connections to facilitate evacuation or sheltering, compared to owners. As a result of these differences, responses to disasters may be quite disparate. Research demonstrates that vulnerabilities can affect responses at every phase of a disaster, including:

• Preparedness. Actions undertaken prior to an event, such as disaster planning, having supplies on hand, securing the home and contents, and installing window protection that can reduce or eliminate potential impacts. Although minority status and lower-income are associated with higher risk perceptions for hazards, minorities and low-income households usually display lower levels of preparedness.

• Warning. Disaster warning processes begin with receiving and then believing a warning, where source credibility and confirmation can be critical, and hopefully end with undertaking protective action such as evacuating. Findings suggest that minorities may experience potential delays in receiving and confirming warning messages since they display greater dependence on informal social and familial networks.

• Evacuation. Research on evacuation is somewhat equivocal, but has found that minorities, lower-income groups, and the elderly are less likely to evacuate, and when they do, they tend to leave later.

• Casualties and Damage. Minorities and low-income groups are much more likely to be disproportionately impacted and hence more vulnerable to disasters. In large measure, this appears to be due to trickle-down housing processes in the United States whereby the poor and minorities are often relegated to older and poorer qual-

ity housing and segregated into less desirable and potentially more risky neighborhoods.

• Reconstruction and Recovery. Minorities, low-income households, and even female-headed households can be at a disadvantage in part because of low language skills and education when it comes to qualifying for and negotiating the process of obtaining public financial resources. Further, racial/ethnic minority groups are often excluded from community post-disaster planning and recovery activities because they have less economic power and political representation. Finally, poorer households and neighborhoods often fall far short of receiving necessary aid to jump-start the recovery process, particularly when it comes to qualifying for loans and obtaining private insurance settlements necessary for housing recovery.

Social vulnerability factors are important determinants of disaster responses and should be considered when undertaking disaster planning related to warning, response, impact, and recovery. Further, socially vulnerable populations are not evenly distributed throughout communities. Instead, they tend to be clustered in neighborhoods. On one hand, such clustering exacerbates the impact of disasters; on the other hand, it may also make it possible for public officials to address such disparate outcomes through spatially-targeted efforts both prior to and after a disaster. In the next section, we explore the use of a spatial decision-making tool to both identify and address the needs of SV populations.

**Social Vulnerability Mapping: The Coastal Planning Atlas Approach**

The inclusion of SV factors into community planning and vulnerability analysis has been slow to develop. Indeed, it was not until nearly the turn of the twenty-first century that researchers began to call for the systematic application of social vulnerability perspectives at the community level to develop social vulnerability mapping (Morrow 1999). The objective is to identify concentrations of populations with particular SV characteristics and thereby identify neighborhoods that will perhaps require special attention, planning efforts, and mobilization to respond to and recover from disasters and hazards.

Our goal in creating social vulnerability maps is to use readily available data from secondary sources, such as the US Census, to allow for broad application of the technique while still providing for sufficiently fine resolution

so that planners and emergency managers might be able to easily identify homogeneous pockets of socially vulnerable populations. Block-groups offer a viable compromise in that a host of data is available to measure dimensions of SV while also being sufficiently small enough in spatial scale that they often match homogeneous neighborhoods. Table 6.1 below displays the 17 indicators utilized to identify socially vulnerable populations. The indicators include a range of factors related to household structure, age, transportation dependence, housing characteristics, minority status, poverty, educational status, employment status and language skills. Having individual SV dimensions or measures available to map at the local level is beneficial because planners can easily identify and perhaps focus on particular types of vulnerabilities given specific hazard risks (table 6.1).

These basic indicators are combined in the second column to form second order SV measures. These indicate special needs that might be relevant during emergency response or disaster recovery, such as areas with higher potential for child care needs, elder needs, transportation needs, temporary shelter and housing recovery needs, and civic capacity. By adding across all indicators, a composite SV score can be created, indicating hotspot concentrations of social

Table 6.1. Social vulnerability indicators.

| Base Social Vulnerability Indicators (percentages) | | |
|---|---|---|
| | 2nd Order | 3rd Order |
| Single parent households with children/Total households | Child care needs | |
| Population 5 or below/Total population | | |
| Population 65 or above/Total population | Elder care needs | |
| Population 65 or above & below poverty/Pop. 65 or above | | |
| Workers using public transportation/Civilian pop. 16+ and employed | Transportation needs | |
| Occupied housing units without a vehicle/Occupied housing units | | |
| Vacant Housing units/Total housing units | Temporary shelter and housing recovery needs | Socially vulnerable hotspot |
| Persons in renter occupied housing units/Total occupied housing units | | |
| Non-white population/Total population | | |
| Population in group quarters/Total population | | |
| Housing units built 20 years ago/Total housing units | | |
| Mobile Homes/Total housing units | | |
| Persons in poverty/Total population | | |
| Occupied housing units without a telephone/Total occupied housing units | | |
| Population above 25 with less than high school/Total population above 25 | | |
| Population 16+ in labor force and unemployed/ Pop in Labor force 16+ | Civic capacity needs | |
| Population above 5 that speak English not well or not at all/Pop > 5 | | |

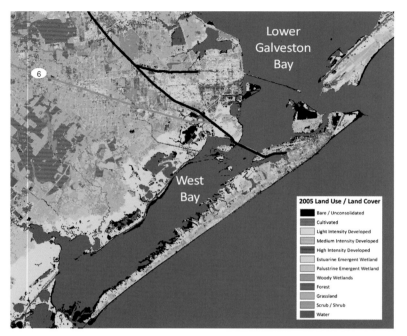

Figure 6.1. Land use and land cover on Galveston Island. The island has a dense urban core at the east end. Photo courtesy NOAA.

vulnerability within and across block-groups in a community.

**Galveston and Social Vulnerability**

Galveston is one of the most urbanized barrier islands in the United States. Although the decline of the population of the city is only partially because of Hurricane Ike (just under 50,000 following the storm), growth in the region has been rapid. Galveston Island has a dense urban core on the east end of the island (fig. 6.1). In 2000, 89 percent of the population lived on the east end of the island. The remain-

ing 11 percent of the population live on sprawling west end of the island and in one additional smaller incorporated community, Jamaica Beach. The urban core has much higher population densities, higher occupancy rates (85 percent, compared to 47 percent on the west end of the island where there are many vacation homes), and higher home ownership rates (60 percent, compared to 46 percent).

Like most cities, housing in the dense urban core is much older and generally in much poorer condition compared with the surrounding area. Not surprisingly, this area also has a much more diverse population, with higher concentrations of minorities and households living in poverty. Figure 6.2 displays block group data for the island urban core, indicating concentrations of minority households. This figure also displays category 1 and 2 surge zones (cross-hatched shaded areas), clearly overlapping a number of block-groups that have relatively high minority concentrations and are also located in areas vulnerable to surge inundation.

The benefit of being able to identify areas that are both physically and socially vulnerable by overlapping these data is identifying critically vulnerable areas to focus on for emergency management and mitigation activities. When Hurricane Ike passed over the island, the urban core was protected from powerful surge flows and destructive wave action coming from the ocean side by Galveston's famous seawall. Nevertheless, rising water entered the urban core from the bay side, flooding category 1 and 2 zones as well as most of the remaining urban core. Category 1 and 2 areas are substantially lower and thus, home and business structures in those areas were subject to extensive flooding prior to the storm passing over the island, making evacuation difficult and time-consuming.

In light of the literature that suggests these populations are less trust-ing of authorities when it comes to heeding warnings and are more dependent on social networks, local emergency management and planning officials might develop special relationships with churches and civic organizations in these areas. These relationships help ensure that when official warnings are released, these organizations can reinforce the warnings through informal networks, enhancing timely compliance. These areas might take priority with urban search and rescue and emergency health officials, as they will quickly visit after a disaster to determine if

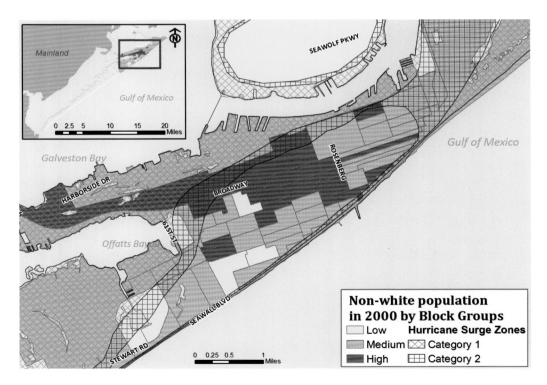

Figure 6.2. Non-white population concentration and category 1 and 2 surge zones (2000).

there are stranded individuals or individuals needing medical attention.

## Findings from Hurricane Ike Research

Hurricane Ike provided an opportunity to assess how well the mapping of social vulnerability characteristics in the *Coastal Atlas* can help assist local emergency management and planning departments identify areas such as neighborhoods containing households and individuals that will have greater difficulty responding to and recovering from a severe storm event like Hurricane Ike. In this section, we present examples of how social vulnerability characteristics are related to the response, impact, access to recovery resources, and the initial stages of the recovery process following Hurricane Ike.

We draw on primary data collected on Galveston Island in the first few months after the storm by researchers at the Hazard Reduction & Recovery Center at Texas A&M University. Unfortunately, these data do not include residents of multi-family structures, which are home to a population that is likely to be particularly vulnerable, since they are almost exclusively renters. As a result, these findings likely underestimate the true incidence of vulnerability among Galveston residents.

Damage assessments identified the structural characteristics of the housing unit as well as visible evidence of damage. Household surveys asked respondents to assess their own level of damage, and also asked a series of questions about evacuation, recovery resources, and early decision-making with regard to returning to the island to rebuild. We also draw on data from the City of Galveston on building permits granted in the months after the storm. These data help us to assess the value of the damage sustained, as well as the timing and volume of repairs undertaken on damaged properties.

A variety of approaches could be used to assess the correspondence or relationships between SV measures and response, impact, and recovery outcomes. A simple but limited approach might be to compare SV maps to outcome maps and look for commonalities in patterns. Another approach would be to calculate correlations between SV indicators and measures of different disaster responses to identify both the direction (positive or negative) and magnitude of the relationship. Here, we use both, drawing conclusions based on the slightly more sophisticated correlations (not shown), but illustrating results using maps and photos.

*Evacuation*

In the hours preceding landfall, residents received continual information related to hurricane warning and watches, and local emergency management (the Galveston County Judge) called for all residents to evacuate the island. The data indicates approximately 80 percent of the population evacuated from Galveston Island. However, neighborhoods with higher percentages of single-parent households, renters, households in poverty, and non-white households experienced lower evacuation rates. Figure 6.3 displays block-group concentrations of households without access to their own vehicles, based on 2000 US Census data. The SV literature suggests that households without access to their own vehicles have greater difficulty evacuating and hence leave later in the process. Figure 6.4 displays average evacuation times by block group from our primary data.

By comparing figs. 6.3 and 6.4, a general overall pattern of correspondence between transportation dependence and late evacuation times can be roughly seen. Not surprisingly, areas with higher concentrations of households without a vehicle also evacuated at lower rates. Many of these same vulnerabilities were associated with later evacuation times. Specifically, neighborhoods with higher proportions of

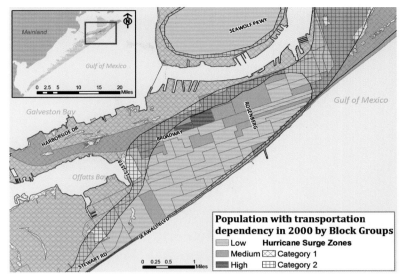

Figure 6.3. Block groups with concentrations of transportation-dependent households and category 1 and 2 surge zones (2000).

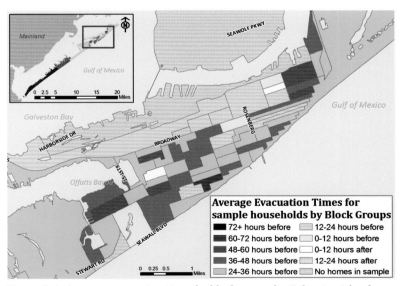

Figure 6.4. Average evacuation times by block group for Galveston Island before Hurricane Ike.

renters, households in poverty, and minorities were more likely to have waited longer to leave Galveston Island in advance of the storm, which greatly jeopardized their evacuation, particularly since surge water began creeping on the island well in advance of the landfall of the storm, cutting off many evacuation routes.

*Damage*

The most visibly devastating impact of the storm is the damage to physical structures, especially homes. As with most hurricanes, the damage comes in two forms: wind and water. In the case of Galveston Island, the wind damage was fairly limited. The real damage came from the storm surge, which washed back across the island from the bay side. The surge that impacted most of the urban core was not the powerful Gulf surge seen on the Bolivar Peninsula, nor was it accompanied by the damaging effects of wave action that destroyed homes and scoured away

whole structures and their foundations. Rather slowly rising waters from the bayside crept on to the island and into the urban core area, filling the city and its homes with water. Consequently, it was difficult to assess water damage from outside the home.

On the whole, the results from Hurricane Ike suggest that the relationship between neighborhood SV and damage was not evident. This finding may well suggest that, at least with respect to damage, SV analysis is of limited utility. However, this non-finding may also be a function of the unique characteristics of Hurricane Ike and Galveston in general. More specifically, this result may be a function of the particular nature of Ike: an extensive but gradual surge flooding event from the bayside of the island with very limited wind damage, and in nature of development on barrier islands, confounding the relationship between physical vulnerability and real estate damage (figs. 6.5 and 6.6).

Oftentimes the literature finding

Figure 6.5. Near the urban core, properties sustained severe flooding as well as surge, which pushed homes off their foundations. Photo courtesy Shannon Van Zandt.

Figure 6.6. A vacation home sits well above the surge level. Only a block off the Gulf Coast, these high-quality homes received only wind damage, which as seen here, was quite minimal. Photo courtesy Shannon Van Zandt.

a relationship between SV and relative damage is based on earthquake and wind-related events in which poor quality housing, generally occupied by SV populations, is shaken apart or picked apart by winds. Furthermore, the work on floods generally is associated with inland communities where low-lying, flood-prone areas have poor land values, and are typically the sites for low-income housing and households. In the case of Hurricane Ike however, we have an event characterized by slowly rising surge that impacted essentially all of the urban core, which is home to 89 percent of the inhabitants of the island, as well as the housing of the relatively affluent on the west end with its outstanding view and proximity to a beach and bay. These latter homes may be at great physical risk, but their owners have little social vulnerability, not only because these homes are often vacation homes and not primary residences, but more importantly because these households typically have very good access to resources—social, physical, and financial—to help them avoid lasting effects from the storm. Thus, the nature of this event and the unique characteristics of Galveston may well account for the lack of relationships between the SV measures and most damage measures.

*Recovery Resources*

Given that damage was widespread and that it affected households in neighborhoods of all income levels and race/ethnicities, one might hope that recovery would also be fairly even and widespread. In this section, we explore the relationship of social vulnerability characteristics with recovery resources. While households may have access to a variety of resources for recovery after a disaster, insurance is perhaps the most commonly accessed, supplemented by public resources from the Federal Emergency Management Administration (FEMA) and other sources in the days and weeks after a disaster.

Our data suggests that neighborhoods with higher proportions of elderly, non-white, and low education households have a greater proportion of residents that are likely to be without homeowner's or renter's insurance. By far the most disturbing finding, given the nature of this disaster, is the result for flood insurance. Neighborhoods with high proportions of minorities and those with higher proportions of individuals that did not complete high school have lower percentages with households covered by flood insurance. These findings suggest that these neighborhoods in particular will be slow to recover because of a lack of private recovery resources (fig. 6.7).

Figure 6.7. In front of this middle-class home, a recreational vehicle provides temporary housing while reconstruction work is completed on the house. Only two months after the storm (November 2008), this middle-class household had relied on private insurance to expedite the recovery process. Photo courtesy Shannon Van Zandt.

We also consider the availability and use of common public resources for recovery aid. There are a variety of forms of more "public" aid with help from FEMA and the Small Business Administration (SBA). Our data indicates that neighborhoods with higher SV indicators also had higher proportions applying for FEMA assistance, yet lower proportions applying for SBA or both FEMA and SBA assistance. Higher density neighborhoods, neighborhoods with higher proportions of single-parent households, households without a car, renters, those in poverty, homes without phones, and the unemployed apply to FEMA only. On the other hand, neighborhoods with higher concentrations of socially vulnerable households have relatively lower applications to the SBA. These findings are not surprising because applications for an SBA low-interest loan suggests the ability to repay that loan, which will be much more difficult in poorer areas and for older individuals who are reluctant to incur increased debt due to their age or financial status.

*Recovery*

Recovery is signified by building activity—home owners, business owners, and residents undertaking repairs to damaged homes or businesses, or rebuilding on their lots after homes have been destroyed. The data suggest that areas with high concentrations of minorities, low education levels, and older housing have been slower to undertake significant repairs/rebuilding, drawn fewer permits, and were later in getting first permits on average. On the whole, the findings with respect to the early stages of recovery clearly suggest that older, more disadvantaged neighborhoods are certainly rebuilding and recovering at a slower pace, and the literature suggests that they are less likely to ever recover. In some cases, these neighborhoods may become targets for redevelopment, meaning the properties are demolished and replaced with different uses, for higher-income housing or nonresidential uses (Yang and Peacock 2010). In cases like these, communities may see an overall loss of affordable housing and may displace original residents, perhaps permanently (figs.6.8 and 6.9).

## Comprehensive Disaster Mitigation and Recovery Planning for Resilience

Resilience implies the ability to resist or absorb impacts and rapidly bounce back from those impacts. In the case of natural disasters and social systems, this implies the ability and capacity to prepare, respond, withstand the disaster impacts without major damage, and most importantly, bounce

Figure 6.8. One of several public housing developments in Galveston, this apartment complex was fenced off almost immediately after the storm. Residents were permitted only a few hours to retrieve their belongings before being permanently removed from the site. These buildings have since been demolished and have not been replaced as of September 2010. Photo courtesy Shannon Van Zandt.

back from the impact sustained. But the picture is often far more complex because communities are composed of networks of businesses, governmental organizations, and households and families living in areas that make up a complex mosaic of socially-defined neighborhoods. They can be substantially different in terms of their socioeconomic composition, the quality and types of housing, and their access and ability to mobilize resources when "bad" things happen. In a very real sense, social vulnerability mapping reveals disparities that make a difference when it comes to the capacity of residents and households to respond, mo-

bilize resources, and bounce back from natural or other types of disasters.

Comparing needs predicted by the *Coastal Atlas* to actual needs expressed after Hurricane Ike, we were able to indeed identify neighborhoods that failed to heed or were slower to respond to calls for evacuation, had lower levels of private and public resources, particularly resources necessary for rebuilding, repairs, and ultimately recovery. Identifying neighborhoods in this way allows community planners, emergency management personnel, and civic leaders to utilize such information to identify neighborhoods where they can work with local civic orga-

Figure 6.9. Two months after the storm, the household living in this home had been unable to clear all their debris or begin repairs on their home. While this home was boarded up, others were left just as they were the day after the storm. Photo courtesy Shannon Van Zandt.

nizations, target education programs, locate emergency shelters, and coordinate evacuation pick-ups to better meet the needs of these populations.

Finally, our approach identified neighborhoods that were on the slow track to recovery and at jeopardy of failing in that pursuit. These failures have consequences not only for the households in those areas, but for the community as a whole, because these become areas at risk of becoming pockets of economic and social despair that can threaten the overall resilience of the community, particularly if they spread. With this approach we can better plan for and monitor our community vulnerabilities and develop more comprehensive planning approaches that can enhance long-term community resiliency.

# 7

# Emergency Management and the Public

*Bill Wheeler*

## Introduction

Over the past 100 years, the speed of communication has increased dramatically, playing a vital role in disaster preparedness, response and recovery. Today, electronic communication is critical to everyday life and the average person is accustomed to having every form of information at their fingertips. In contrast, during the Galveston Hurricane of 1900 critical warnings were communicated via telegraph, a great asset despite the time it took for a message from the National Weather Service (NWS) in Washington to arrive in Galveston. Today, storm communications have advanced to the point where NOAA Hurricane Hunter flights in active storms transmit data that is received and reviewed by the public in real time. With the speed and turnover of different message delivery systems today, the challenge for emergency management in the coming years will be mastering those systems and formatting the messages so that they can be delivered to communities before it is too late.

## Disaster Communication

Robust communication between public officials, emergency managers, and the community is critical for the successful defense and recovery of the Gulf Coast region in the event of a severe storm (fig. 7.1). This communication, tied with public education efforts, gives individuals and communities the tools necessary to plan ahead. To enhance community understanding of the risks of an approaching hurricane, these tools are communicated to the public by the media through public service announcements. In addition, it is necessary for elected officials to routinely discuss disaster plans and information in the media, increasing assurance, guidance, and public response during a disaster event (fig. 7.2). Communication between emergency management personnel and the community is the first line of defense in preparing for a disastrous event and is important for post-event recovery as well.

Every year the National Hurricane Center (NHC) and the NWS launch their annual hurricane preparedness

Figure 7.1. The control room at Houston TranStar, the first of its kind, combines the resources and expertise of the Office of Emergency Management with transportation managers into one state-of-the-art center that monitors traffic and road closures. TranStar is a partnership of four government agencies that are responsible for providing transportation management and emergency management services to the Greater Houston region. Photo courtesy Houston TranStar.

Figure 7.2. Harris County Judge Robert Eckles, FEMA FCO Tom Costello, and Senator Joe Lieberman of Connecticut talk with FEMA staffers at the Houston Disaster Recovery Center (DRC). The DRC is comprised of federal, state and local agencies. Photo courtesy Ed Edahl/FEMA.

campaign. At the local level communication in the event of a severe storm lies with the senior elected official, the Director of Emergency Management. In Texas, and many other states, that responsibility belongs to the County Judge. It is important, however, that all senior elected officials along the coast cooperate to make sure the affected populations are informed of the risk of an approaching storm and make the necessary provisions in preparing for, responding to, and recovering from a disaster event.

The emergency management responsibility of each senior elected official is a very large task. In most coastal jurisdictions, this responsibility has been placed with a professional Emergency Management Coordinator (EMC). Supporting the EMC is a staff of trained professionals who develop plans to address the threat of an approaching storm. These plans include communication with the community in order to convey seasonal preparedness themes, warning messages, and post-event recovery. The format and functionality of communication is handled by the Public Information Officer (PIO). Communication during an event is routinely reviewed and approved by the Incident Commander (IC).

## Communicating with the Public

Since hurricanes Katrina and Rita in 2005, it has been recognized that robust preparedness actions at a local level are the first defense. Both local and federal governments have focused on citizen preparedness and the jurisdictional message has been increasingly aimed at the individual. This message can range anywhere from "Hurricane season has arrived: Keep your tanks filled" to "Turn around! Don't drown!" (fig. 7.3). To make sure the message is communicated to constituents and that it reaches everyone in the community, emergency managers plan public information programs prior to each hurricane season. These plans include public service announcements, commercial media purchases, media events by commercial outlets, community presentations, and workshops with direct public contact.

Every year, US coastal communities prepare for June 1, the first day of hurricane season. This is a good time for each jurisdiction to begin disseminating a seasonal message. Leveraging the message with the media coverage of the year's hurricane season is a winning combination. However, the media hype surrounding the beginning of hurricane season varies from year to year and can be measured by the previous hurricane season. For example, if the community was greatly affected by a storm, then the media coverage will be quite lengthy. On the other hand, if the community has not been affected in a number of years, dissemination of hurricane preparedness information may be quite difficult. In most communities where storms have not had an effect in a significant number of seasons, media coverage is minimal at the beginning of each season.

Any communication plan should function 24 hours, 7 days a week and be capable of providing vital information up to 30 days after a disaster, well into the recovery period. The messages will be vetted by the organizations PIO and approved by the IC of the jurisdiction prior to distribution. Continuity of the message is very important. As the communication plan is developed, common sub-themes should remain consistent throughout. For example, evacuation terms used in the pre-event messages should remain the same in the event of a storm. Similarly, the message during the event should reflect the annual message. It is very important that emergency preparedness terms are explained well and it may be necessary to define complex terms to the public. It is important to disseminate the message early and repeat it often to make sure it is reaching at-risk communities.

Inviting media partners to view the communication plan and familiarize themselves with the format and theme

is very important. The media is the key partner and is the best way to disseminate information to the at-risk community. The relationship between emergency managers and the media can be complemented with scheduled workshops, news conferences, and press re-

Figure 7.3. In recent years, community awareness signs have been placed in flood-prone areas and at low-water crossings, advising drivers to avoid high water. Photo courtesy NOAA.

leases. By developing a well-rounded relationship in this matter, the media will count on emergency management officials for the most up-to-date information and provide it to the community, minimizing false information and increasing valuable response. Emergency management officials continually search for the most efficient way to communicate with the public, both before and after disaster events (fig. 7.4). Although media venues reach most of the community, there are areas of society that are harder to reach, such as elderly and low-income communities. However, using the vast electronic and social media methods in existence today, communication with that sector of the community should become easier.

## Pre-Event Communication: Evacuation

During the approach of a hurricane, most people want to know where the hurricane will make landfall. Although this piece of information is good to know, it is not the most important. Most people do not recognize that their perceived risk and actual risk may not be equivalent. Gulf Coast emergency managers should identify common risks for their populations and attempt to successfully communicate with the at-risk population. For example, advising a community to weather-in-place when their risk is low will leave more space on the roads for more seriously at-risk communities. On the other hand, if a community is in danger of

Figure 7.4. Harris County Judge Ed Emmett announces Houston TranStar's plan for evacuees with special needs. The assistance plan encourages those with special needs, including the disabled and elderly, to register with the county for transportation assistance during a mandatory evacuation. Photo courtesy Houston TranStar.

flooding from storm surge and has a very large special needs population (e.g., retirement home), the community should be advised to evacuate.

One of the most valuable and yet the most difficult programs to deploy is an evacuation assistance plan for special needs members in a community (fig. 7.5). This program offers transportation assistance to persons living in a mandatory evacuation area to reduce their risk of drowning. The program is open to those needing transportation assistance or those who need help due to illness or mobility concerns. Unfortunately, the availability of this program has been communicated to the community and the agencies that

serve this population with only marginal success. The primary reason for this lack of success is that the special needs population is very suspicious about giving up personal information. In the Houston/Galveston area, public education programs have lowered these barriers and most eligible citizens have begun signing up for evacuation assistance prior to hurricane season.

The success of this program will ensure that emergency managers can efficiently evacuate the vulnerable populations from hurricane surge zones in a timely manner before the onset of tropical storm force winds. In recent years, deploying multiple methods has

Figure 7.5. Evacuees with special needs wait to be bused to Austin during Hurricane Ike. Photo courtesy Mayra Beltran/ *Houston Chronicle.*

been successful in encouraging registration. The program managers have also maintained contact with these individuals once they are registered in the program and continually check on the need to maintain the service to those individuals. The communication with these individuals is clear, as the message explains the risk and the danger of not registering, effectively increasing registration.

One of the most difficult decisions for an individual to make is whether to evacuate before an approaching storm. In order to make a good decision, emergency managers need to communicate the correct risk information. Putting the risk information in plain English is difficult, but can be done with the right communications plan. If the citizen understands their risk in the event of an approaching hurricane, they can make an informed decision to stay and weather-in-place. If the message conveys concern and increased risk is conveyed, then the citizen may make an informed decision to evacuate. It is important to keep in mind that the tone of the message will play a huge role in the decision of a citizen to evacuate.

**Event Communication**

After landfall, the most difficult issue to address is how to keep the community listening. The only solution is to make a plan early and execute according to the plan. In a community that has been recently effected by a severe storm event, it is easy for the public to remember how information was distributed. It is in those communities that have not been impacted in a number of years where the greatest challenges exist. In some places, communities have never experienced a hurricane and many emergency plans have never been tested. In these situations, it is important to identify those communication tools that can be easily adapted and used by the public to gather important information to make critical decisions.

One of the most valuable tools used in recent years and in the aftermath of hurricanes Katrina, Rita, and Ike was simulcast television media over the radio. Despite widespread power outages, many people had access to a car radio to get critical information. These tools proved invaluable in the first hours after Hurricane Ike. Battery-powered televisions were also very useful, however digital television has compromised the use of analog television. It is important to keep in mind that new means of communication are difficult for the older population because they do not routinely use new communication tools. This population stays connected by telephone and television and will not be able to navigate new informational paths during a disaster.

Once the community has identified ways to communicate effectively without normal utility power, those means should be publicized and tested to increase awareness about how to gather important information during an event. Presenting the complete picture to the public helps the public deal with the disaster and process the information to start the recovery process. It is important to dispel any rumors that may be crisscrossing the community. Keeping the public informed with the facts allows them to concentrate on next steps for their family and the community as well as not wasting valuable resources on rumors.

Communities that have executed a Disaster Communications Plan effectively all report that communication with the public must continue during and after the event and that the message must be conveyed using the same methods to which the community has become accustomed. Governmental agencies in charge of preparing for, responding to, and recovering from hurricane disasters must make plans well in advance and consistently communicate a clear message on how the public should expect receive emergency communications. Local governmental agencies must partner with their media outlets to make these communications pathways easy to start up and operate so there is seamless communication with the public.

**Post-Event Communication**

After the event, it is very important that clear, factual information is communicated to the public as reports from damage assessment teams become available (fig. 7.6). Reasonable care should be given to making sure the information about the status of the affected community is timely. Rumors should be addressed quickly and replaced with the facts about the real issues. Often, depending on the severity of the incident, facts relating to death and destruction must be communicated. That should be done in a timely manner and with careful consideration of those affected by the loss. Communication, although directed at the adults handling the response and recovery after the event, is often comprehended by children. This time can

Figure 7.6. Texas Governor Rick Perry provides a public update on the status of Hurricane Ike. Photo by Adam Beaugh, courtesy the Office of the Governor.

Figure 7.7. Thousands of Hurricane Katrina survivors from New Orleans settle into a Red Cross shelter in the Houston Astrodome. At least 25,000 evacuees, mainly from the Louisiana Superdome, lived in the Astrodome until September 17, 2005. Photo courtesy Andrea Booher/FEMA.

be very traumatic for children and a confident, caring communications plan that considers a child's fears can make a long-term difference in how children recover from the disaster event. Most are already traumatized, but calm, clear communication will have a better outcome and speed the emotional recovery process.

The evacuated population should not be forgotten. Communicating with that population may be difficult; they are often sheltered far away (fig. 7.7). The PIO should make working and corresponding with the sheltering community and evacuated population a high priority. The sheltering community will also be interested in the progress toward stabilizing the jurisdiction so they can plan for their eventual return.

Once the damage has been communicated to the public, it is important to disseminate information about the steps to recovery. The jurisdictions should explain what the community should expect in every aspect of the recovery process. For example, in communities affected by major hurricanes, citizens should be briefed on debris management, distribution of commercial commodities, supplies that have been disrupted, power restoration, and the eventual opening of Disaster Recovery Centers (DRC) (fig. 7.8). This message will have to be presented many times on multiple days and in many formats through all operating media outlets to make sure the message makes it to all areas of the affected community. When the community becomes aware of the situation, the recovery process has begun (fig. 7.9).

## The Message Does Not Change

In the 1960s, US citizens were educated by Civil Defense authorities and the message was communicated over and over again. During that time, hurricane preparedness was taught in Houston schools. The message was clear and those that experienced it still remember the message today. Today, given the frequency of major incidents, it is important to make sure the community understands the message and can respond should a disaster strike. It is not the government that will be the first responder, it is the individual that is at the epicenter of the event.

Figure 7.8. An aerial view of the North Carolina Baptist Men's disaster kitchens Manna 1 (left center), Unit 2 (right), and Unit 3 (left), a FEMA commodity Point of Distribution (POD) (upper left center), and a Red Cross command center. The three kitchen complex could cook and distribute 70,000 meals per day; 2 weeks after Hurricane Ike and this kitchen group was still sending out 7000 meals per day. Photo courtesy Mike Moore/FEMA.

Figure 7.9. Residents are allowed entry into Galveston Island nearly two weeks after Hurricane Ike made landfall in the area. Traffic backed up miles out of town as residents made their way back to see their houses and begin cleanup. Photo courtesy Jocelyn Augustino/FEMA.

A successful response is dependent on the dissemination of the preparedness message to the community so that citizens can make plans for all types of hazards. The community that prepares the best will recover much faster than communities that have barely prepared. As communities experience disasters, they learn how to handle the events, and find better ways to prepare and respond. From these lessons learned, the message must be adjusted to make response better and more efficient. Emergency plans have evolved

and improvements have been seen in hurricane evacuations across Florida and Texas, communications between first responders during the World Trade Center attack in New York City, and national preparedness programs, like Citizens Corp, that educate the community. All these improvement processes make communities more prepared and allow citizens to be self-sufficient until the response and recovery begins.

Communication of the important messages on preparedness is the most important day-to-day public information an emergency management organization can deliver. There are a variety of communication tools that the public uses on a daily basis and that message must be pushed every day. Billboards, radio, television, and print are some of the most widely used methods of disseminating the preparedness message in the community, but as technology expands, the information must get pushed out using new tools such as social networking media. The greatest challenge for public information today is disseminating the personal preparedness message to all segments of the population through the correct communication technology. The second challenge is to keep the message fresh and confident and to keep from spreading fear throughout the community. Finally, the message must be kept in front of the community year-round. The following chapter discusses emergency evacuation and transportation planning in the Houston/Galveston area.

# 8

# Emergency Evacuation and Transportation Planning

*Carol Abel Lewis*

## Introduction

On June 1 each year, Gulf Coast residents begin watching storm systems cross the Caribbean and listen as meteorologists predict approaching tropical cyclones. Given the forecast of the strength and point of landfall of a storm, emergency management officials determine the appropriate response and implement emergency procedures. In most cases, there is a call for a voluntary or mandatory evacuation to clear residents out of sensitive or low-lying areas. A key component of hurricane and disaster planning is moving people out of locations where dangerous conditions and potential loss of life are likely. Rapid population growth in the Gulf Coast amplifies the need for discussion about how to safely and efficiently evacuate these areas in the face of imminent storms. More than 700,000 people are projected to move into the evacuation zones by 2035 (HGAC 2007). However, future expansions of the roadway system are projected to be minimal, exacerbating a well-known, critical roadway capacity issue.

In 2005, Hurricane Rita was predicted to make landfall in the Houston/Galveston area. After watching the devastation in New Orleans caused by Hurricane Katrina, many residents evacuated earlier than recommended and ignored suggestions to "shelter-in-place." Fear led to extreme congestion across the city as residents watched their neighbors leave and made decisions to evacuate. A mass exodus ensued in the region. Evacuees reported a typical 40-minute trip at 8 or more hours and the normal 4-hour travel time to Dallas at 20 hours (fig. 8.1). Hurricane Rita eventually

Figure 8.1. Traveling along I-45N, Houston based evacuees leave mandatory evacuation areas and head north. Photo courtesy Smiley Pool/*Houston Chronicle*.

made landfall near the Texas/Louisiana border, bringing very little rain or wind to the Houston/Galveston area. However, Rita indirectly caused over 100 deaths due to traffic accidents and heat exhaustion.

In contrast to Rita, evacuation during Hurricane Ike in 2008 progressed very smoothly and emergency professionals and residents generally considered it a success. Two key changes were implemented after Rita that helped greatly improve evacuation during Ike. First, improved communication between elected officials reduced constituent confusion. Second, applying zip codes to the recommended zonal map more clearly instructed residents when they should leave. The following discussion explains the basics of the transportation system and evacuation plan, offers insight into mobility during an evacuation and presents considerations for future evacuation planning in Texas Gulf Coastal areas.

## Houston Galveston Area Transportation System and Evacuation Plan

As discussed in chapter 7, in Texas, decisions to evacuate are made by local officials. Print, auditory, and visual media caution people against complacency during the hurricane season and encourage preparation and family pre-planning. Local governmental agencies and the state have websites with extensive information providing guidance about what to do before, during, and after a hurricane. Officials sponsor community seminars that assist residents in making plans. Families are encouraged to gather their emergency kits and consider whether they will shelter-in-place or evacuate. If evacuating, citizens are advised to pre-determine their travel route (fig. 8.2).

All levels of government are involved in successfully handling an evacuation situation. State governors, emergency management offices, state transportation officials, and local governments are extremely visible and work together to achieve the most streamlined outcomes. Behind the scenes, public safety personnel, transportation engineers and planners, and police officers conduct assessments and make recommendations that affect the flow of evacuation traffic. In non-emergency times, government planning organizations, private consulting firms, and university researchers spend thousands of hours collecting background information that helps to facilitate smoother responses during evacuations. This includes the analyses of the routes available, capacity to handle traffic volumes, and remedies for anticipated or unanticipated hindrances or bottlenecks.

During an evacuation, four Houston area freeways constitute the preferred routes for moving residents out of the region: two head north (I-45, US 59) and two northwest and west (US 290 and I-10 west, respectively) (fig. 8.3). It is recommended that evacuees head north or west in the event of a hurricane, choosing either I-45 toward Dallas, US 290 toward Austin, or I-10 West toward San Antonio. Traveling east on I-10 is not advised since that portion of the interstate heads southeasterly and is not sufficiently inland to serve as a safe haven. US 59 North is excluded as an evacuation route if the Beaumont area is also evacuating.

These interstate and US highways are supported by state and local roadways; especially important are SH 146, SH 6 and Beltway 8 which connect the coastal evacuation zones to the primary highways and interstates. To alleviate potential bottlenecks and facilitate traffic flow, certain major evacuation routes are suggested to the public. These routes are determined from observations made during previous evacuations. Computer modeling can also indicate where severe problems may occur.

During Hurricane Rita, when mass evacuation took place, transportation professionals were criticized for not immediately implementing contraflow as part of the emergency evacuation plan. Contraflow allows emergency

personnel to reverse inbound freeway lanes, creating additional capacity (fig. 8.4a, b). Shortly after Rita, TxDOT created a contraflow plan. Contraflow capability is now available for I-10 West, I-45, US 290, and US 59 North allowing emergency personnel to open both directions of traffic at choke points to relieve congestion and improve travel speeds at crucial locations. During Hurricane Ike, engineers did not need the full contraflow plan, but found that the strategy improved travel flow for a small section of I-45 north, where the freeway narrows to two lanes.

The Federal Emergency Management Agency (FEMA) defines sev-

Figure 8.2. A sign points the way for the hurricane evacuation route in Texas. Photo courtesy Jocelyn Augustino/ FEMA.

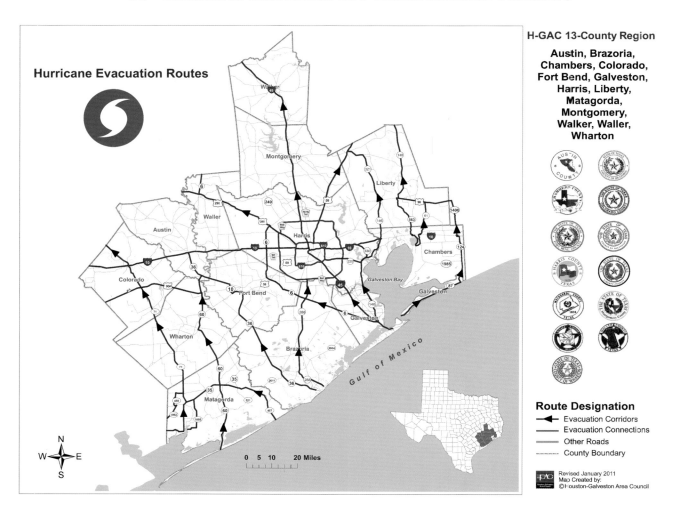

Figure 8.3. Hurricane evacuation routes for the Houston/Galveston region. Courtesy of Houston/Galveston Area Council.

eral categories of evacuation. Evacuation for events that are spontaneous, or no-notice events, is less applicable during a severe storm event because it would be too late to utilize prediction. Fortunately, hurricane tracking by the National Weather Service allows emergency management officials to determine appropriate actions hours in advance. In the event of a hurricane, three types of evacuation may take place: mandatory or directed evacuation, voluntary evacuation, and shelter-in-place.

*Mandatory or Directed Evacuation.* This is a warning to persons within a designated area that an imminent threat to life and property exists and that individuals *must* evacuate in accordance with the instructions of local officials (FEMA 1996). *Voluntary Evacuation.* This is a warning to persons within a designated area that a threat to life and property exists or is likely to exist in the immediate future. Individuals issued such a warning are not required to evacuate; however, it would be to their advantage to do so. *Shelter-in-Place.* In the event of an impending hurricane, individuals and families may assess their own level of potential discomfort because of an approaching storm. In this case, officials consider the risk of life-threatening danger as relatively low.

Residents in the Houston/Galveston area are advised about the evacuation status of their home or business using a color-coded zip-zone map (fig. 8.5). The four zip zones (Coastal, A, B and C) are grouped by zip code, reducing confusion about who should evacuate. The evacuation area extends roughly 120 miles in length and 40 miles across from Matagorda County to Chambers County. Parts of Brazoria, Galveston, and Harris counties are included within that expanse. Zip-Zone Coastal and Zip-Zone A are yellow; they are most prone to storm surge

Figure 8.4a. A sign designating when and where to begin contraflow during a hurricane evacuation. Photo courtesy Houston TranStar.

**(a) Normal Operations**

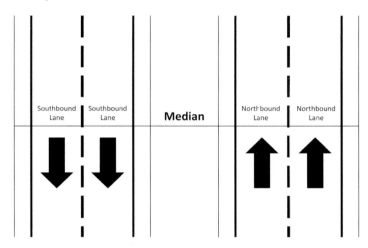

**(b) Normal Plus Two Contraflow Lanes**

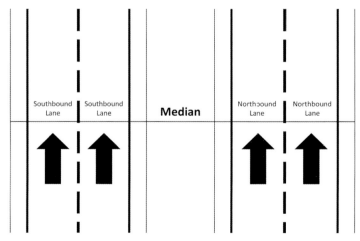

Figure 8.4b. Depiction of traffic flow when contraflow is allowed. Courtesy Rice University Archives.

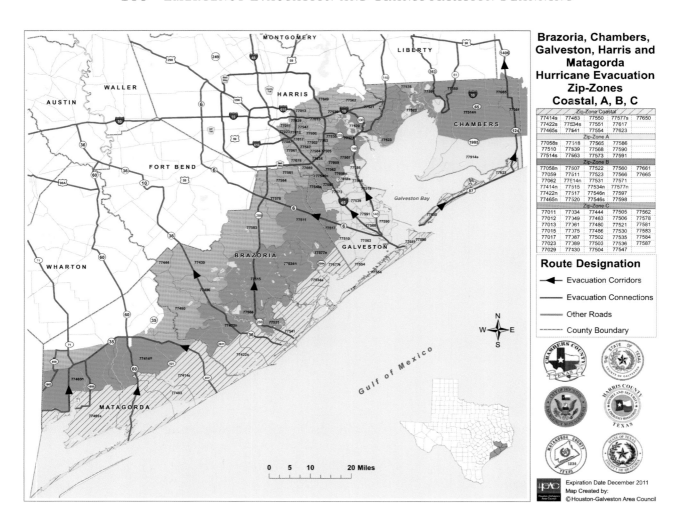

Brazoria, Chambers, Galveston, Harris and Matagorda Hurricane Evacuation Zip-Zones Coastal, A, B, C

| Zip-Zone Coastal | | | | |
|---|---|---|---|---|
| 77414s | 77483 | 77550 | 77577s | 77650 |
| 77422s | 77534s | 77551 | 77617 | |
| 77465s | 77541 | 77554 | 77623 | |

| Zip-Zone A | | | |
|---|---|---|---|
| 77058s | 77318 | 77565 | 77586 |
| 77510 | 77539 | 77568 | 77590 |
| 77514s | 77563 | 77573 | 77591 |

| Zip-Zone B | | | | |
|---|---|---|---|---|
| 77058n | 77507 | 77522 | 77560 | 77661 |
| 77059 | 77511 | 77523 | 77566 | 77665 |
| 77062 | 77514n | 77531 | 77571 | |
| 77414n | 77515 | 77534n | 77577n | |
| 77422n | 77517 | 77546n | 77597 | |
| 77465n | 77520 | 77546s | 77598 | |

| Zip-Zone C | | | | |
|---|---|---|---|---|
| 77011 | 77334 | 77444 | 77505 | 77562 |
| 77012 | 77349 | 77463 | 77506 | 77578 |
| 77013 | 77361 | 77480 | 77521 | 77581 |
| 77015 | 77375 | 77486 | 77530 | 77583 |
| 77017 | 77387 | 77502 | 77535 | 77584 |
| 77023 | 77389 | 77503 | 77536 | 77587 |
| 77029 | 77430 | 77504 | 77547 | |

**Route Designation**

➤ Evacuation Corridors
— Evacuation Connections
— Other Roads
---- County Boundary

Expiration Date December 2011
Map Created by:
©Houston-Galveston Area Council

Figure 8.5. The hurricane evacuation zip-zones map helps to identify which areas are under a mandatory evacuation and which should hunker down during a hurricane. Courtesy of Houston/Galveston Area Council.

and are the first identified to evacuate. Zip-Zone B (green) is the second tier and Zip-Zone C (orange) is the most inland with sections extending into the east side of Houston. Portions of some zip codes are within 5 miles of downtown Houston. NASA is located in Zone A and much of the Port of Houston Authority, one of the largest ports in the

nation, is located in Zones B and C.

**Roadway Traffic Volumes during an Evacuation**

Residents and decision-makers agree that evacuation prior to Hurricane Ike proceeded better than with Hurricane Rita. A number of task forces, including one formed by Governor Rick Perry after the 2005 Hurricane Rita evacuation, attempted to delineate what went wrong and determined necessary adjustments. The task force named five areas that required attention and better structure: articulation of command and control, fuel availability, traffic management, attention to persons with special needs, and public outreach and more coordinated communication (Little 2006).

Based on these recommendations and subsequent procedural adjustments, command and control occurred more smoothly during Hurricane Ike and information provided to the citizenry was coordinated between elected officials. Maps with zip code designations clarified evacuation messages, and citizens adhered more closely to the zones. Vehicles left the region less haphazardly.

During Hurricane Rita when the mandatory order was issued for Zone A, people throughout the Houston/Galveston area began leaving, clogging routes that lead out of the region. Residents in many mandatory areas are known to have turned back due to the extreme congestion. During Ike, however, clearing the coastal areas at greatest risk for storm surge was aided by lower than typical traffic on designated evacuation roadways since many employers closed businesses and schools dismissed roughly 36 hours prior to the storm's landfall. Houston Mayor Bill White requested the following of employers the day before Hurricane Ike made landfall: "We ask employers, in order to keep their employees safe and community safe and our freeways clear of traffic, to take into account those recommendations and only require employees in essential services to show up to work. Do not penalize employees who are not in those needed positions who do not show up to work" (Texas Cable News, September 12, 2008).

This action removed a number of vehicles from the roadways that would have been making home-to-work and other business-oriented travel while the evacuation began. Removing many non-essential inland trips increased available capacity for persons in the mandatory evacuation zones.

In addition to improvements aiding traffic flow, arrangements were made with fuel providers to ensure that there would be sufficient supplies of gasoline to meet the needs of travelers. Furthermore, fewer people in voluntary evacuation zones chose to leave the region during Hurricane Ike than Hurricane Rita. Prior to Rita, an estimated 2–3.5 million persons left Houston (fig. 8.6). During Hurricane Ike, the total number was estimated to be about 1.2 million persons. An online voluntary survey conducted by TranStar, the Houston region's joint governmental emergency management organization, indicated that

Figure 8.6. Traffic congestion along I-45N during Hurricane Rita required authorities to open contraflow lanes, allowing one-way traffic on all parts of the freeway. Wikipedia at http://en.wikipeedia.org/WiKi/File: Rita-HoustonEvacution.jpg

77 percent of respondents did not leave during the Ike (TranStar 2009). Harris County's highest ranking official, Judge Ed Emmett, spoke often of "hunkering down" in communications with area residents. Clearly, the reduction in volume, combined with the communication and traffic management improvements, resulted in more efficient evacuation and contributed to better evacuation flow.

## Future of Evacuation

The population in Zip Zones Coastal, A, B, and C is projected to increase more than 60 percent from 2005 numbers (table 8.1). According to the Houston-Galveston Area Council (HGAC) 2035

Regional Transportation Plan, 51 lanes are available to evacuate the region (HGAC 2007). More than 415,000 vehicles can be accommodated and an additional 90,000 vehicles may be accommodated if nine contraflow lanes are opened. The agency estimates that it would take 36 hours to evacuate 1 million residents under perfect conditions. This statement alarms professionals, who must plan for future Gulf Coast evacuations.

Perfect evacuation conditions are improbable and resident evacuations will never be ideal. Individuals waiting on relatives and making last-minute decisions about whether to evacuate or shelter-in-place will affect the timing and flow of vehicles on the

Table 8.1. Population projections for zip-zones Coastal, A, B, and C (HGAC 2007).

| ZONE | 2005 Population | 2035 Population Projection | % Increase |
|---|---|---|---|
| A | 83,403 | 139,225 | 67% |
| B | 350,408 | 572,866 | 63% |
| C | 780,425 | 1,279,279 | 64% |
| **Total** | 1,214,236 | 1,991,370 | 64% |

roadways. Furthermore, resident responses to each hurricane will be different and difficult to predict (Peacock et. al. 2007).

Volumes and trip distribution are impacted by variables such as time chosen to leave, number of voluntary evacuees, routes, and ultimate destinations. This complicated variable set combined with higher numbers of people evacuating will result in a fragile balance that must be struck in order to achieve perfect conditions for future evacuations. Murray-Tuite and Mahmassani (2004) concluded that trips made to collect children and conduct other pre-evacuation needs (termed trip-chaining) impact evacuation travel patterns. In addition, traffic demand estimates should be increased 150 percent if the trip-chaining was not previously considered. The point is that the number of evacuees on the road combined with persons making other trips in preparation to evacuate creates additional pressure for the available roadway space.

Area evacuation routes accommodated Hurricane Ike reasonably well with the regional population in Harris and Galveston counties slightly exceeding 4 million people and 1.2 million evacuees. Regional growth forecasts for the next 25 years predict a 75 percent increase of 3 million additional residents (for a total of approximately 7 million) with a planned 36 percent increase in roadway miles as shown in table 8.2. Of that planned roadway increase, some sections of I-45 south leading from Galveston are scheduled for widening to ten lanes. State Highway 146 and SH 6 are slated for minor improvements over the next 25 years, but capacity will not be substantially increased.

Table 8.2. Houston Galveston Area Council 2035 Plan: total regional roadway miles.

| Year | Miles* | Change |
|---|---|---|
| 2007 | 23,520 | - |
| 2025 | 31,916 | 36% |
| 2035 | 32,046 | 0.4% |

*Freeway/tollways, arterials, collectors, and managed lanes

To date, Texas and Gulf Coast hurricane planning discussions have centered on evacuation with little attention given to other options. The nearly 2 million people forecasted to live in zip-zones Coastal, A, B, and C by 2035 make the discussion of additional options an unavoidable component of future regional hurricane planning. As the number of coastal and regional residents increase, it will require more time to clear the at-risk populations. Evacuating 2 million people will more than double the HGAC 36 hour estimate to more than 72 hours, given less than perfect conditions and layering the pre-evacuation trip making previously described by Murray-Tuite and Mahmassani (2004). Beginning evacuation so many hours in advance is precarious because the path and intensity of the storm is so much less certain at that time.

Given the growth projections and lack of expansion in future roadway capacity, it is imperative that elected and emergency officials augment the potential responses to imminent hurricane predictions. Considerations may apply to the entire region, but should focus on zip-zones Coastal, A, B and C. Discussions may include, but are not limited to the following:

*Additional Bus and/or Rail Evacuation.* Evacuate a greater number of individuals by bus or by adding a rail option on the railroad right-of-way which parallels SH 3. Riders from Galveston and surrounding communities could be transported north using bus or rail, reducing pressure on the roadways. The Regional Gulf Coast Rail District is examining commuter rail to Galveston, which may facilitate this as an option. However, challenges include funding and motivating residents to leave their vehicles behind.

*Shelter Nearby in Secure Facility.* Several keynote speakers representing Gulf Coast states at the February 2010 National Evacuation Conference in New Orleans noted a shift in emphasis in their states toward providing local emergency public shelters. The benefit of sheltering nearby is that residents remain closer to home, accelerating the beginnings of recovery. In addition, other cities, counties or states do not have to offer their hospitality. Demand for roadway space is reduced. Among the disadvantages are cost of construction, resident inconvenience and discomfort at mass sheltering, and potential public sector liability.

*Florida has the most* extensive public shelter program in the nation with the 67-county listing of the sites easily accessible on the Internet. The locations report the number of people that can be housed, generally in the thousands, with a companion column showing an assessment of excess or deficient capacity for their county (Florida Disaster 2010). Mississippi (http://www.msema.org/saferooms/index.html) and Alabama (http://www.prlog.org/10598998-fema-grant-funds-chickasaw-community-safe-room.html) created safe room programs designed to accommodate persons during weather events. The rooms can be small and oriented towards individuals or families in their homes or they can be as large as 3200 square feet to shelter large groups.

*Intensify Building Codes to Enable More People to Shelter-in-Place.* A review of existing building codes could be initiated to ascertain whether residential construction in certain areas could be strengthened to endure higher velocity winds. For example, Florida has codes specifically for High Velocity Hurricane Zones for those areas at high storm risk (Uniform Florida Code 2009).

*Limit Construction in Certain Coastal Areas.* Limiting or prohibiting construction in certain coastal areas is the most controversial option in the items for public discourse. Implications for public purchase of some lands, and questions of property rights would require review, assessment and legal queries to determine feasibility. This option is discussed further in chapters 11 and 12.

It is likely, given population projections, that a storm in 2035 will require advances in all areas listed above, in addition to sufficient population evacuation. Beginning public dialogue about additional options offers the opportunity to examine the advantages and disadvantages of each strategy and estimate the number of individuals that could be reasonably accommodated using each approach. The critical point is that projected population growth and available highway capacity will not allow evacuation as a stand-alone response to hurricanes in the future. Discussion by land owners, planners, engineers, and public officials will enable the region to better prepare for the inevitable future storm.

# 9

# Lessons in Bridge Infrastructure Vulnerability

*Jamie E. Padgett and Matthew Stearns*

## Introduction

The performance of regional bridge infrastructure has a significant impact on the safety, effectiveness, and efficiency of the transportation system following hurricane events, which is crucial to facilitating post-event response and recovery activities (Fig. 9.1). Hurricane Ike caused notable damage to the infrastructure of the Houston/Galveston Area when it made landfall on September 13, 2008. Many local bridges were completely destroyed and although the majority of these were small timber structures in rural areas, multiple major bridge structures also suffered damage from debris, storm surge and wave loading. Much of the damage can be attributed to inundation of the decks, or *superstructures,* of the bridges, debris impact, and erosion of abutment supports and approaches.

This chapter presents a holistic overview of the damage to bridge infrastructure in the Houston/Galveston area caused by Hurricane Ike. Typical failure modes are evaluated by assimilating a rich data set of post-event assessment surveys and inspection reports. The data assembled include field reconnaissance conducted by the authors, HNTB (a nationwide bridge design firm) through the Texas Department of Rural Affairs, the Texas Department of Transportation, and interviews with local municipalities or other bridge owners. The performance of timber bridges. often located in rural areas, as well as major highway bridges is assessed. The damage summaries presented include a discussion of factors contributing to the damage, repair procedures, and simple capacity/demand checks for case studies in which bridges over water crossings were damaged during Ike.

Assimilation and assessment of the empirical data from such natural disasters provides key lessons for improving the performance of infrastructure given potential exposure to future events of an even greater magnitude. The chapter concludes with a summary of lessons learned, recommendations for mitigating future damage, and needed future studies that can benefit from this investigation.

Figure 9.1. The I-10 highway bridge near Pass Christian, Mississippi was seriously damaged by the storm surge associated with Hurricane Katrina that lifted sections of the bridge up and off. Photo courtesy John Fleck/ FEMA.

## Damage to Bridge Infrastructure

Both timber structures as well as major highway bridges are considered in this reconnaissance study. Much of the damage experienced can be attributed to storm surge and wave loading on the bridge superstructures. The damage or destruction of 26 structures has been ascribed to these loads. In order to find the levels of storm surge and wave heights experienced by areas along the Gulf of Mexico in the Houston/Galveston Area, SWAN and AD-CIRC modeling was conducted at the University of Texas at Austin (Dawson and Proft 2010). This modeling was discussed in chapter 5. The models revealed that the peak storm surge level was in excess of 14 feet in some locations, while wave heights reached values of over 5 feet near Rollover Pass. The hindcast data was further used to identify the surge and wave height at locations where bridge damage occurred, and was evaluated and compared relative to the hazard estimates.

Twenty-five bridge structures were subject to scour around the abutments and wing walls of the bridges, and four local bridges suffered from impact damage caused by debris. The impact damage caused spalling, or the chipping off of portions of concrete. Damage to surveyed bridges is summarized in table 9.1. The bridge type noted in the table refers to the superstructure and substructure material of the bridge. The state of damage is characterized by the level of damage incurred and is assigned based on the definitions provided in table 9.2. For example, visible, repairable damage to the structure that does not affect the structural integrity is deemed *light damage.*

Using the data from the Hurricane Ike SWAN and ADCIRC simulations, the maximum storm surge and wave heights were found along the coast and plotted in fig. 9.2 relative to the bridges and their respective damage states. The figure excludes 17 rural bridges located further inland that also experienced damages from Hurricane Ike but were not located in the surge zone. These bridges were likely damaged by heavy rains that accompanied the storm as it moved further inland.

The following sections provide a summary of the findings of the reconnaissance data accompanied by a detailed discussion of these failure modes for different bridge types. To define the terminology, fig. 9.3 shows a general bridge layout showing the key components and types of damage typical of bridges studied. The figure illustrates the mean water level prior to the storm, the level of the storm surge, and the level of the accompanying waves in relation to the structure. This increased volume of water can cause deck uplift, scour, or impact damage, as illustrated in the figure. The additional details below provide insight on mitigating such damage in future events.

**Damage to Timber and Local Bridges**

A large majority of the damaged bridge structures during Hurricane Ike were small local and timber bridges. These structures suffered from erosion around the substructure due to the storm surge and increased floodwaters flowing under the bridges. They also suffered impact damage from floating debris. Many post-event inspection reports for the rural structures also noted the deteriorated condition of the bridges. This deterioration cannot be precisely attributed to the storm event and was likely pre-existing; therefore it is not included in the report. However, it is acknowledged that such pre-event deterioration could render structures more susceptible to damage in extreme events such as hurricanes. Hurricane Ike caused a wide range of structural damage, destroying 24 bridges and leaving 18 with heavy damage. The paragraphs below describe the failure modes common for timber bridges and other small local bridge infrastructure.

One of the most common forms of damage to local bridges in the area was scour. The storm surge, flooding, and increased water levels and flow

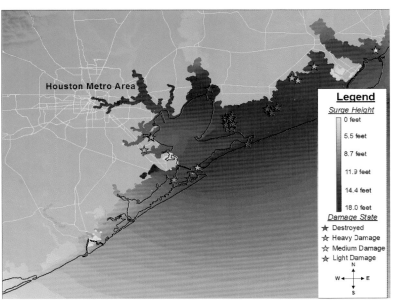

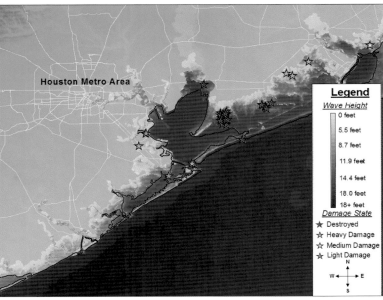

Figure 9.2. Maximum inundation caused by storm surge height during Hurricane Ike relative to damaged bridges in the Houston/Galveston area.

**Table 9.1. Summary of damaged bridges in Houston/Galveston area.**

| County | Street Name | Damage State | Damage Description | Type | Spans |
|--------|-------------|--------------|--------------------|------|-------|
| Chambers | Cain Bayou #1 | Destroyed | Deck unseated | Timber | 2 |
| Chambers | Cain Bayou #2 | Destroyed | Deck unseated | Timber | 2 |
| Chambers | Cain Bayou #3 | Destroyed | Deck unseated | Timber | 2 |
| Chambers | Cain Bayou #4 | Destroyed | Deck unseated | Timber | 3 |
| Chambers | Edwards #1E | Destroyed | Deck unseated | Timber | 3 |
| Chambers | Edwards #1W | Destroyed | Deck unseated | Timber | 4 |
| Chambers | Edwards #2E | Destroyed | Deck unseated | Timber | 3 |
| Chambers | Edwards #2W | Destroyed | Deck unseated | Timber | 3 |
| Chambers | Edwards #4W | Destroyed | Deck unseated | Timber | 3 |
| Chambers | Fitzgerald #1 | Destroyed | Deck unseated | Timber | 3 |
| Chambers | Fitzgerald #2 | Destroyed | Deck unseated | Timber | 4 |
| Chambers | Fitzgerald #3 | Destroyed | Deck unseated | Timber | 2 |
| Chambers | Fitzgerald #4 | Destroyed | Deck unseated | Timber | 2 |
| Chambers | Jay Matthews | Heavy Damage | Scour damage | Timber | 3 |
| Chambers | Lake Anahuac #1 | Destroyed | Deck unseated | Timber | 2 |
| Chambers | Lake Anahuac #2 | Destroyed | Deck unseated | Timber | 2 |
| Chambers | Leyroy Easer #1 | Destroyed | Deck unseated | Timber | 4 |
| Chambers | Leyroy Easer #2 | Destroyed | Deck unseated | Timber | 8 |
| Chambers | Spencer #1 | Destroyed | Deck unseated | Timber | 2 |
| Chambers | Wanda Lagow #1 | Destroyed | Deck unseated | Timber | 3 |
| Chambers | Wayne Morris #1 | Destroyed | Deck unseated | Timber | 3 |
| Chambers | Wayne Morris #2 | Destroyed | Deck unseated | Timber | 2 |
| Chambers | Wayne Morris #3 | Destroyed | Deck unseated | Timber | 2 |
| Chambers | Wayne Morris #4 | Destroyed | Deck unseated | Timber | 4 |
| Galveston | 6th St. Bridge (East) | Light Damage | Scour damage | Concrete | 2 |
| Galveston | 7th St. Bridge | Destroyed | Deck unseated | Timber | 4 |
| Galveston | Clear Lake Dr. | Light Damage | Scour and spalling damage | Concrete | 2 |

**Table 9.1, cont'd.**

| County | Street Name | Damage State | Damage Description | Type | Spans |
|--------|-------------|--------------|--------------------|------|-------|
| Galveston | Pelican Island Bridge | Destroyed | Scour damage | Concrete | 92 |
| Galveston | Rodeo Bend | Heavy Damage | Scour damage | Concrete | 1 |
| Galveston | Rollover Pass Bridge | Destroyed | Deck unseated | Concrete | 5 |
| Galveston | Tiki Drive | Light Damage | Spalling damage | Concrete | 3 |
| Jefferson | Boondocks Bridge | Heavy Damage | Scour damage | Steel | 6 |
| Jefferson | Craigen Bridge | Heavy Damage | Scour damage | Steel | 5 |
| Jefferson | Humble Camp Bridge at Hildebrant Bayou | Heavy Damage | Deck displaced | Concrete, Steel, Timber | 24 |
| Liberty | Nueces Road | Heavy Damage | Scour damage | Concrete | 2 |
| Liberty | Rice Belt Road (CR105) | Medium Damage | Scour damage | Timber | 2 |
| Liberty | River Road | Heavy Damage | Scour damage | Timber | 1 |
| Orange | E. Roundbunch Rd. | Medium Damage | Scour damage, electrical failure | Concrete | 6 |
| Polk | Freeman Rd. Bridge | Medium Damage | Impact damage | Steel, Timber | 3 |
| Polk | Union Springs Rd. | Light Damage | Impact damage | Timber | 3 |
| Tyler | CR-2150 Big Cypress Creek Bridge | Heavy Damage | Scour damage | Timber | 2 |
| Tyler | CR-2670 #2 | Heavy Damage | Scour damage | Timber | 1 |
| Tyler | CR-3400 | Heavy Damage | Scour damage, impact damage | Timber | 1 |
| Tyler | CR-3430 at Sugar Creek | Heavy Damage | Scour damage | Timber | 1 |
| Tyler | CR-3625 Ebenezer Church Rd. | Heavy Damage | Scour damage | Timber | 2 |
| Tyler | CR-3630 #1 | Heavy Damage | Scour damage | Timber | 2 |
| Tyler | CR-3630 #2 | Heavy Damage | Scour damage | Timber | 1 |

112

**Table 9.1, cont'd.**

| County | Street Name | Damage State | Damage Description | Type | Spans |
|---|---|---|---|---|---|
| Tyler | CR-3725 | Medium Damage | Scour damage | Timber | 2 |
| Tyler | CR-3725 at Pamplin Creek | Heavy Damage | Scour damage, impact damage | Timber | 3 |
| Tyler | CR-4600 Pump Station Rd. | Heavy Damage | Scour damage | Timber | 5 |
| Tyler | CR-4825 Hester Rd. | Heavy Damage | Scour damage | Timber | 4 |
| Tyler | CR-4875 Midway Rd. | Heavy Damage | Scour damage, impact damage | Timber | 4 |

**Table 9.2. Damage state labels and descriptions.**

| Damage State | Description |
|---|---|
| Light | Some reparable damages to the superstructure. No immediate danger. |
| Medium | Minor damage to the superstructure and possibly substructure of the bridge. Possible loss of structural integrity. |
| Heavy | Major damage to entire bridge structure. Severe loss of structural integrity, posing public danger. |
| Destroyed | Bridge structure unusable or missing. |

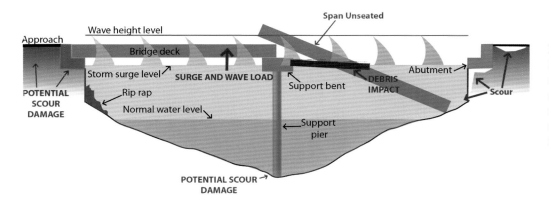

Figure 9.3. Sample bridge during hurricane event showing key structural components, regions susceptible to damage, and common failure modes.

often led to deterioration of riprap by the abutments, undermining of bridge approaches, and damage to the areas behind the wing walls of many structures. The Ebenezer Church Road Bridge experienced scouring behind the abutment wing wall, where the majority of the material behind the wall was swept away (fig. 9.4). The storm surge and floodwaters from Hurricane Ike also carried large debris that caused a significant amount of damage to local bridges. Many bridges had debris resting against the superstructure which affected post-event functionality or caused visible structural damage due to debris impacts. The Union Springs Road Bridge is an example of a structure with impact damage. During the storm a tree had fallen onto the bridge and was found resting on the middle support bent.

When storm surge and waves accompanying storm events such as Hurricane Ike rise at or above the level of the bottom of the bridge deck, the deck is subjected to uplift and transverse forces that can cause significant damage to the structure. If the surge and waves are high enough and there

Figure 9.4. The Ebenezer Church Road Bridge suffered erosion behind the abutment wing wall during Hurricane Ike. Photo courtesy HNTB Corporation.

is limited connection between the bridge superstructure and substructure, this loading can shift or even completely displace the deck of the bridge. The 7th Street Bridge was subjected to storm surge loading, and the deck was displaced completely off the support structure (fig. 9.5). Such damage was frequently observed in rural timber structures after Hurricane Ike.

## Damage to Major Bridge Structures

Much of the damage afflicting smaller local bridges also occurred in major bridge structures and roadways. For the purpose of this chapter, bridges with more than five spans constructed of steel and concrete are considered

Figure 9.5. The 7th Street Bridge experienced deck unseating from storm surge loading during Hurricane Ike. Photo courtesy HNTB Corporation.

major bridge structures. Three bridges will be used to demonstrate the damage to major structures in the Houston/ Galveston Area during Hurricane Ike: Humble Camp Bridge at Hildebrandt Bayou, Rollover Pass Bridge, and Pelican Island Bridge (fig. 9.6). These cases offer examples of various bridge failure modes during hurricanes, and insight into design details that affect the performance of bridge infrastructure subjected to hurricane induced loads.

The Humble Camp Bridge crosses Hildebrandt Bayou is a concrete and timber bridge with 24 spans; there are nine concrete and steel spans on the northwest end of the bridge and the remaining 15 spans are timber and concrete (State of Texas 2009). During the storm, a 13-foot surge coupled with 3.3-foot waves overtopped the bridge causing the timber-supported spans to shift slightly in the transverse direction. This indicates a potential failure of the vertical and transverse restraints in the bridge, but not complete failure of the structure or loss of the superstructure.

The bridge was built with shear key, or shear links, in the form of steel plates bolted between the bridge deck and bent beam (fig. 9.7a, b). As visible in the picture, multiple spans have shifted as a result of the storm surge wave loading and the shear keys of the bridge bent on some spans but did not fail completely. It is likely

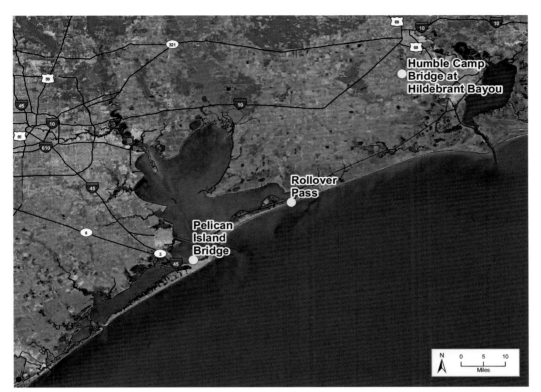

Figure 9.6. The locations of Pelican Island Bridge, Rollover Pass Bridge, and Humble Camp Bridge at Hildebrandt Bayou are shown in this image. These three bridges suffered major damage during Hurricane Ike. Courtesy Rice University Archives.

that these shear key details helped the bridge remain intact without the span unseating.

The Rollover Pass Bridge carries State Highway 87 across Rollover Pass on Bolivar Peninsula. It was one of the most severely damaged bridge structures during Hurricane Ike. Rollover Pass Bridge was a concrete box-girder bridge with five spans on a concrete pile substructure. It had a deck clearance of approximately 5.3 feet over mean water elevation. During the storm the bridge was subjected to a 15-foot surge with 5-foot waves. Many of the sections of the bridge suffered major shifting and others were removed completely, resulting in the displacement of four of the five spans. Site visits indicated that a small amount of spalling occurred on the bent beams along with complete loss of most connections between superstructure and substructure. Overall, the abutments and support bents of the bridge suffered little to no damage, according to post-event inspection reports (Merrill, personal communication, January 15, 2010), and were salvaged for the repair and restoration of the bridge.

Figure 9.7. Shifted spans of the Humble Camp Bridge (a) and close-up of shear key (b) after Hurricane Ike.

Figure 9.8 shows the shifting and unseating of the spans and their box-girder subcomponents under the surge and wave loading. The shifting and unseating can be attributed primarily to limited continuity and connectivity. Adjacent bridge spans responded in a non-continuous manner and box-girders of the same span were shifted independently. In many cases, bridge spans shift as a single unit or resist surge wave loading. However, the post tensioning cables between adjacent girders failed under the hurricane loads allowing individual girders to shift. It is worth noting that corrosion of these cables was visible from field assessment. Furthermore, only limited connectivity was provided between the superstructure and substructure by dowels that suffered pullout. Figure 9.9 shows the extent of the damage to the structure, including the missing and shifted spans and failed connections. After the storm, one lane was opened, restricting traffic while emergency repairs were conducted using a combination of new and salvaged girders to restore the functionality of the bridge.

The roadway adjacent to the Rollover Pass Bridge, State Highway 87, in Bolivar also suffered extensive damage from inundation and flooding due to storm surge. The Texas Department of Transportation (TxDOT) reported that the road had many sections that were undermined by the water and

Figure 9.8. Hurricane storm surge uplifted sections of Rollover Pass Bridge on Bolivar Island. Photo © 2011 Bryan Carlile, Beck Geodetix.

completely swept away. TxDOT used local supplies of milled asphalt to fill in the damaged areas to allow transportation of goods over the roadway following the event. To restore functionality, it was also necessary to remove an extensive amount of debris, including approximately 10,000 pounds of hazardous materials, 250 cars, boats, and golf carts, and over 40,000 cubic yards of sand that had been distributed across the highway by the surge (K. Kipp, personal communication, February 23, 2010).

The Pelican Island Bridge is a concrete slab bridge with a movable bascule section in the main span to allow for ship navigation. The majority of the damage to the bridge was to the approaches, according to post-event inspection reports (L. Jarosz, personal communication, February 16, 2010). The concrete slab portion of the bridge superstructure did not suffer any damage during Hurricane Ike. The bascule section of the bridge had severe damage to the fender systems, which were in need of repair, as they offered minimal protection to either side of the bridge.

Site visits and reports from Sullivan Brothers Builders also revealed that the Southwest approach of the bridge was undermined by erosion (fig. 9.10a). Sections of the approaching roadway and the rip rap that was

Figure 9.9. Damage to Rollover Pass Bridge (a), including missing girders and pullout of dowels and superstructure-substructure connection (b). Photo courtesy Brian D. Merrill, TxDOT.

designed to protect the roadway were also eroded (L. Jarosz, personal communication, February 16, 2010). This erosion completely exposed the tops of the abutment cap (fig. 9.10b). Severe erosion also occurred around the support columns of the concrete section of the bridge, amplifying the already-present erosion problem (Sullivan Brothers, personal communication, June 3, 2010). After the event, emergency repairs to the approach were completed in six days by filling in the scoured areas to allow traffic to pass over the bridge. These initial repairs cost $400,000, and TxDOT bid out the final repairs to bring the approach back to its original condition.

**Lessons Learned**

The post-event reconnaissance and assessment conducted in this study reveals that a combination of hurricane-induced flood and surge as well as wave action resulted in observable damage to many bridges during Hurricane Ike. The rich data set assembled indicates that approximately 53 bridges in the area sustained damage, and although many of these structures were rural timber bridges, significant damage was also inflicted to concrete and steel bridge structures. Historically, limited data regarding performance of these bridge types has been available in the literature. However, the extensive damage during Ike revealed the significant vulnerability of timber structures to extreme storm events and demonstrates the need for their retrofit or replacement to enhance safety in a region susceptible to hurricanes.

In addition to washout of timber structures and other rural bridges, Hurricane Ike also provided lessons regarding the performance of typical highway bridges with steel or concrete superstructures, including major water crossings. Several of these findings are consistent with the empirical evidence of failure modes and potentially improved design details revealed in past hurricane events. These include damage attributed to storm surge and wave loading, impact from debris, and scour or erosion of abutments and foundations. Given the potential for hurricanes with even higher storm surge across the Houston/Galveston area depending upon the strength of storm and angle of incidence along the coast, these lessons could be critical considerations as investments are made to upgrade or replace bridge infrastructure in the region.

Many of the reports issued by the Texas Department of Rural Affairs (TDRA) indicate that the observed timber and other rural structures require complete replacement, suggesting new concrete structures with increased

**a**

**b**

Figure 9.10. eroded roadway on the Pelican Island Bridge approach (a) and exposed abutment of the bridge (b). Photo courtesy Sullivan Land Services, LTD.

clearance above storm surge and wave action and increased structural capacity. Other alternatives to replacement may also exist, however, such as capacity enhancement of existing structures or partial replacement of select components. For example, using grated decks on bridges may help to alleviate the surge and wave loading on superstructures which is common during hurricanes and reduce the exposed surface area. This option to mitigate the forces of storms on bridge superstructures may be warranted for non-timber structures as well. Many of the rural timber bridges could be replaced with box culvert structures rather than traditional bridge structures, since they typically span very small water crossings. Approach mitigation or erosion backfill may still be required. Other retrofit options to enhance the capacity of these bridge types are fairly similar to the non-timber structures discussed below. However, all of these partial replacement and retrofit options warrant further investigation and are possible areas for future research.

Lessons regarding failure modes of bridge infrastructure during Hurricane Ike are often consistent with the assessment of performance in such past storm events as hurricanes Katrina and Ivan. Many bridges severely damaged during Katrina were uplifted from their support, similar to the damage observed during Ike at Rollover Pass and many of the rural timber structures (DesRoches et al. 2006; Mosqueda et al. 2007; Padgett et al. 2008; Okeil and Cai 2008). All of these bridges are water crossings that have multiple span superstructures with low elevation, little or no continuity between adjacent spans, and limited connectivity to the substructure.

For new structures, continuous superstructures with increased clearance levels over the water would be obvious solutions to further protect the structure from damage during extreme storm events. However, for retrofit of existing structures or design of approach spans that tie into the roadway, simple design details such as transverse shear keys or restrainer cables could be viable options. In fact, the avoidance of unseating on the Humble Camp Bridge during Hurricane Ike can be partially attributed to the shear links provided between the deck and piers. However, care should be taken to avoid transferring forces in excess of the pier or foundation capacity, or these elements must also be strengthened. Alternatively, a fuse in the strengthened connection between superstructure and substructure could be provided so that uplift of the deck is avoided in moderate events up until the force transferred to the substructure exceeds a fraction of the substructure capacity. At that point, deck unseating would be deemed an acceptable alternative as long as the costly pier and foundation substructure is salvaged. While this alternative is not ideal to promote immediate post-event functionality in extreme events, it may mitigate the cost of repair in severe cases.

In addition to improved structural details, the reconnaissance after Hurricane Ike revealed the need to mitigate susceptibility to scour and erosion at the approaches, abutments, and foundations. This was previously demonstrated by damage from Hurricane Ivan. Improved abutment wing walls or riprap at the pier foundations and abutments may be used to protect the structures from scour or erosion damage, because many of the existing wing walls are either constructed of timber or nonexistent. Additionally, land barriers can also serve as protection to approach spans. For example, reports and site visits revealed that the only area of the Pelican Island Bridge approach that was not severely damaged in Hurricane Ike was protected by a thin plot of land that extends into Galveston Bay. This small amount of extra buffering was enough to protect that section of the approach from any erosion damage while sections on both sides were completely destroyed.

## Conclusion

Hurricane Ike was a devastating storm for the Houston/Galveston area that provided a number of lessons regarding the vulnerability of bridge infrastructure and potentially viable improvements in design details for coastal bridges nationwide. Similarities in failure modes sustained by these structures have been observed in past storm events, however the data assimilated here provides some of the best empirical evidence of bridge infrastructure performance for both major bridges and rural structures relative to estimates of the hurricane hazard at the bridge sites. Factors contributing to bridge damage are highlighted and can be used to help screen vulnerable or high priority bridges. While a number of options for improved design details are supported by empirical evidence, further research is required to analytically and experimentally assess the viability of many of these strategies. The continued development of improved retrofit and design details, coupled with the ongoing development of reliability-based vulnerability models (Padgett et al. 2009), will help to improve the safety of bridge infrastructure in hurricane exposed regions.

# 10

# Hurricane Impacts on Critical Infrastructures

*Hanadi S. Rifai*

## Introduction

Hurricanes have far-reaching impacts at numerous levels on the natural and built landscape. The effects are felt for years and the recovery process can be long and slow depending on the nature of the event. This chapter focuses on the impacts of hurricanes on critical infrastructure along the Texas Gulf Coast, specifically on the impact of Hurricane Ike on the Houston/Galveston area. Critical infrastructure is defined in this chapter in the broadest sense to include the built landscape, municipal infrastructure, environment, natural resources, industry (including ports and shipping), power and transportation networks, and public and private services. In addition to describing damage from Ike, this chapter provides some insight into potential damage to the municipal and petrochemical sectors in the event of a larger storm.

Hurricane Ike affected the regional infrastructure to varying degrees. Ike caused catastrophic damages to Galveston and coastal resources in Texas, but further inland, the most

severe effects were loss of power due to toppled power poles and lines and damage on residential, commercial, and industrial structures caused by fallen trees. Some of the effects of the storm were secondary impacts due to primary effects on another element. For example, the integrity of water distribution network was lost due to loss of power. Fewer damages were incurred from flooding when compared to surge and high winds.

## Built Landscape

Urban built landscapes are sophisticated and complex systems. They include commercial and residential built space with interconnected municipal services such as water, sewer, power, storm water, telecommunications, fuel, and solid waste facilities. Hurricanes have multi-faceted effects on this infrastructure ranging from destruction to disruption of services or damage to specific components. In general, there are three types of possible damages: wind damage, storm surge damage, and flood damage. Wind damage can destroy poorly constructed buildings and send debris such as roofing materials, signs, and siding flying. Storm surge is capable of removing a home from its foundation and completely destroying relatively weak structures. Floodwater and surge can inundate structures causing water damage and conditions suitable for the development of mold. Vehicles left in residential neighborhoods are typically submerged and destroyed.

Commercial infrastructure includes stores, buildings, and commercial strip centers, whereas service infrastructure includes schools, public buildings, power transmission networks, and gas stations, among others. Hurricanes significantly impact this sector, through physical destruction and damage as well as by interrupting services. For example, in schools in Galveston, classrooms and common facilities were destroyed by high winds and water damage. Thousands of children were unable to return to school because class was canceled for almost a month while facilities were repaired.

During Hurricane Ike, thousands of homes along the upper Texas Gulf Coast were severely damaged (fig. 10.1). Bridge City and Gilchrist are two communities that were hard hit by Ike and had all but 15 homes within their residential neighborhoods completely destroyed (FEMA 2008). Waterfront

cities between Galveston and Houston sustained residential infrastructure damage (e.g., Dickinson, Kemah, League City, and Clear Lake Shores in Galveston County, and El Lago, La Porte, Nassau Bay, and Seabrook in Harris County), but more importantly, had substantial losses in their sub-regional economies which rely on tourism, water recreation, NASA, and the Houston Ship Channel industry. Table 10.1 displays the real estate losses by dollar amount of damage after Ike. Upwards of 3300 units were destroyed and more than 100,000 units experienced damage amounting to more than $25,000 per unit.

**Power Loss**

Hurricane winds and storm surge can damage power networks and cause power outages to significant areas. Power grids are fragile systems with centralized control and no redundancy in sub-transmission and distribution paths. Power outages are associated with numerous secondary effects that may be more costly than the primary loss of power. They plunge areas into darkness and idle pumping stations, cutting off water supplies and allowing wastewater release. During Ike, more than 2 million residents of the greater Houston area lost power (fig. 10.2) and it took 19 days for normal service

**Table 10.1. Number of eligible real property losses by damage level since December 3, 2008 (from FEMA 2008).**

| County | Eligible Real Property Losses (Owners) by Amount of Damage | | | | Number of Owners Affected | Number of Properties Destroyed |
|---|---|---|---|---|---|---|
| | Up to $8000 | $8001–$15,000 | $15,000–$28,000 | >$28,000 | | |
| Orange | 3,656 | 781 | 2,254 | 986 | 7,677 | 99 |
| Harris | 56,583 | 1,933 | 1,417 | 609 | 60,542 | 482 |
| Galveston | 11,420 | 2,427 | 3,495 | 2,232 | 19,574 | 2,228 |
| Chambers | 2,130 | 181 | 128 | 249 | 2,688 | 345 |
| Jefferson | 11,277 | 306 | 320 | 223 | 12,126 | 218 |
| Total | 85,066 | 5,628 | 7,614 | 4,299 | 102,607 | 3,372 |

to return to most residents. Power outages in the Houston area were primarily caused by downed trees and power lines due to high winds (110 mph). In Galveston, storm surge flooded four of the power substations. Thus, the majority of the damage was in the distribution network, rather than in transmission and substation power infrastructure. There have been proposals to install underground power transmission; however, it should be noted that such systems have a significantly higher cost and greater operational limitations. They cost approximately 2–4 times more than overhead lines, and take substantially longer to repair since they are buried underground. Additionally, lines cannot be repaired "live" (while the circuit is functional) due to their proximity to earth. Thus, underground power lines are usually installed in densely populated and sensitive areas, such as the Texas Medical Center in Houston.

**Municipal Infrastructure**

One of the critical impacts of hurricanes that affect communities collectively is the interruption and damage to the water and sewer infrastructure. Most municipalities have sophisticated networks of drinking water plants, wastewater plants, pump stations, water towers, collection and distribution networks, and discharge pipes that release their effluent to streams and rivers. In addition to the possibility of destruction and inundation during a hurricane, this infrastructure is vulnerable to loss of power, pipe breakage, backflow, and cross-flow contamination. These problems occur in both water and wastewater infrastructure. Backflow is essentially the unwanted reversal of flow caused by gravity, vacuum, or other pressure changes in the sewer lines. Cross-flow contamination occurs when wastewater enters and contaminates the water supply system via a cross connection between the two systems. In water systems, backflow causes downstream materials to enter the water supply system, cross-contaminating the water. In a wastewater system, backflow due to heavy rain can cause fluids to back up into homes or flow out of manholes.

As was the case during Ike, power loss can occur without any secondary source for power generation, causing pressure loss in pipe networks. The safety of the water supply system becomes compromised. The loss of power during Ike also caused unintended releases of effluent from wastewater plants into their receiving streams. Some water and wastewater plants in affected areas did not have backup power generation systems, while others had systems that did not function. Residents experienced low water pressure in their homes and in some instances their water flows were down to a trickle for up to three days after the storm. Furthermore, the government mandated water boiling to ensure the

Figure 10.1. Two aerial photographs taken six days apart illustrate the destruction of residential structures and beach erosion on Bolivar Peninsula after Hurricane Ike. Photo courtesy US Geological Survey.

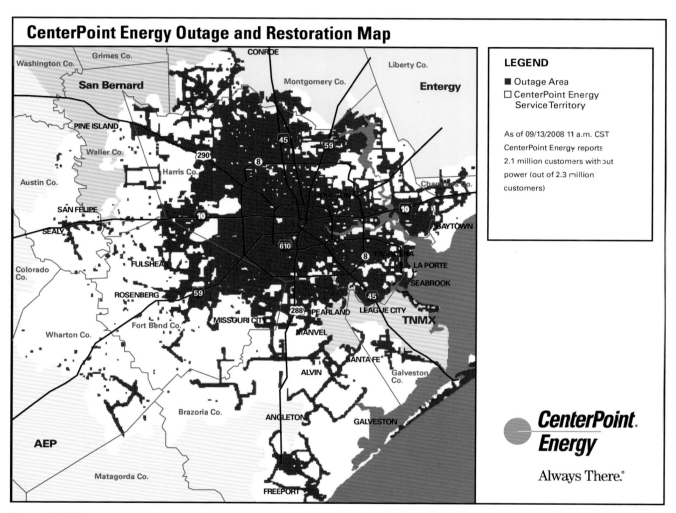

## CenterPoint Energy Outage and Restoration Map

**LEGEND**
- ■ Outage Area
- □ CenterPoint Energy Service Territory

As of 09/13/2008 11 a.m. CST CenterPoint Energy reports 2.1 million customers without power (out of 2.3 million customers)

Figure 10.2. Power outage in the greater Houston metropolitan area and Galveston following Hurricane Ike. Courtesy of CenterPoint Energy.

safety of the water. In these situations, typical contaminants that pose a threat to human health are microbial and chemical contaminants. Acute health effects from pathogens such as *Giardia* and *E. coli* often involve gastrointestinal disorders, fever, nausea, and diarrhea.

During Hurricane Ike, the waste-water treatment plant in Galveston was inundated by surge waters and lost power completely. The plant is now being rebuilt and will be elevated by 13 feet. Unlike in coastal areas, wastewater plants in the central Houston area were not exposed to surge. Damage to the water and sewer network was limited to power outages and disruption, and by late November 2008 all systems were operating as usual (FEMA 2008).

While the municipal infrastructure was mostly spared during Ike, the potential for inundation exists if an Ike-like storm were to make landfall further to the west along the Gulf Coast or from an Ike-like storm that has higher wind velocities. Figures 10.3a–c illustrate the potential inundation for selected wastewater facilities in the Houston Ship Channel area. As seen in fig. 10.3a, the facilities (delineated using orange highlights) are outside the 100-year floodplain and were not inundated during Ike. However, in the two scenarios shown in fig. 10.3b and 10.3c the two facilities are inundated. In fig. 10.3b. the ADCIRC predictive model, developed by researchers at the University of Texas at Austin, shows partial inundation of the facilities due to an Ike-like storm with 30 percent higher wind speeds. All of the facilities are under water in fig. 10.3c which shows the modeling results from an Ike-like storm with 30 percent higher wind speeds making landfall further west around Jamaica Beach on Galveston Island.

One of the most severe consequences of hurricanes is the generation of large volumes of debris. The debris is composed mostly of downed trees and vegetation, destroyed housing and commercial buildings, dam-

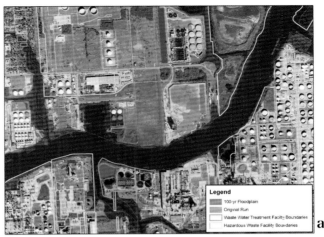

Figure 10.3a. The inundation of critical facilities in the Houston Ship Channel area during Hurricane Ike (modeled using ADCIRC).

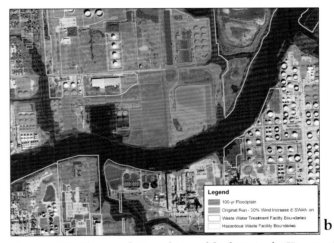

Figure 10.3b. The inundation of critical facilities in the Houston Ship Channel area had winds been 30 percent higher during Hurricane Ike (modeled using ADCIRC).

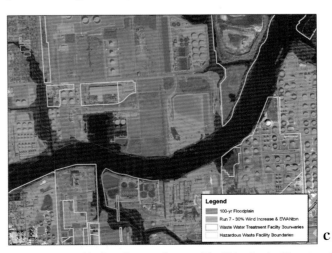

Figure 10.3c. The inundation of critical facilities in the Houston Ship Channel area had winds been 30 percent higher and had the hurricane made landfall near Jamaica Beach (modeled using ADCIRC).

aged vehicles, and a wide array of other detritus containing petroleum, toxic chemicals, plastics, and other materials that may become toxic if burned. Municipalities are generally unprepared to dispose of these kinds of materials. After Hurricane Ike, the total amount of debris was estimated at 25 million cubic yards (approximately 5000 football fields piled a yard high) and the cleanup costs were more than $80 million. The City of Houston was overwhelmed by the need to collect downed trees, and the amount of mulch that was generated from tree disposal. Additionally, some debris was carried by Gulf of Mexico currents hundreds of miles and deposited in areas as far away as Padre Island. The debris impacted national seashores and wildlife refuges potentially endangering wildlife in these protected areas. Debris that did not float in Gulf waters was submerged or semi-submerged in East Bay and Trinity Bay, presenting a threat to boaters in the area for months following Ike. The Texas General Land Office (GLO) conducted a survey of the bay, logging locations of submerged debris and cleaning up debris in the Gulf system.

## Industrial Infrastructure

The oil and natural gas and petrochemical infrastructure is often affected by hurricanes, particularly in the Gulf Coast region. The oil and natural gas sector has production, transmission, and distribution systems throughout the Gulf of Mexico and East Coast regions, and the petrochemical sector is concentrated in the Houston Ship Channel area. The petrochemical sector also has a significant number of offshore facilities, including production facilities and pipelines. In preparation of a tropical cyclone, almost all natural gas production and processing capacity in the Gulf Coast region is shut in. Shut-ins can continue for an extended period of time following a tropical cyclone.

In 2008, there were 55 major natural gas processing plants (more than 150 million cubic feet) in Texas, Louisiana, Mississippi, and Alabama in the path of hurricanes Gustav and Ike (EIA 2008). These plants account for about 38 percent of US processing capacity. Twenty-eight pipelines that move natural gas from offshore production platforms to onshore processing plants were shut down and over 2000 oil and natural gas production platforms (of the 3800 in the Gulf of Mexico) were exposed to hurricane conditions from the two storms. Sixty platforms were destroyed.

Oil and natural gas that is produced in the Gulf of Mexico is transported through more than 33,000 miles of pipelines linking 3800 production platforms. The network includes sub-sea pipes, valves, compressors, and dehydration and separation facilities. Hurricanes can cause significant damage to this network. For example, hurricanes Ike and Gustav broke mainlines causing extensive damage that required time-consuming repairs.

Natural gas processing plants that remove contaminants, extract nitrogen, demethanize, and extract/fractionate natural gas liquids are also affected during tropical cyclones. During hurricanes Ike and Gustav shutdowns occurred because of the lack of electric power, disruption of upstream supplies, inaccessibility of plants because of road conditions, and the inability to send out natural gas liquids from the plants. In addition, plant-specific damages occurred such as flooding, debris damage, and equipment destruction.

The petrochemical industry also sustained significant damage during Hurricane Ike (fig. 10.4a). Oil refineries in Orange and Jefferson counties sustained more damage and flooding than those to the south. Eight chemical companies had 4–10 feet of salt water inside their plants. It should be noted, however, that for the most part where Ike made landfall and its somewhat lower than expected surge levels limited the amount of damage that was incurred by the petrochemical sector. Modeling scenarios using the University of Texas ADCIRC model

show that surge might have been 4 feet higher had Ike landfall to the west of Galveston Island or if wind speeds had been at least 30 percent higher. Figure 10.4b illustrates the additional inundated areas for selected petrochemical facilities in the Houston Ship Channel given a scenario in which Ike is routed to make landfall further west along Galveston Island.

**Transportation Networks**

Transportation networks are also affected by tropical cyclones. The typical impacts include roadway inundation and the destruction of roadways and bridges (due to high winds, surge, and floodwater) as discussed in chapter 9. However, other transportation elements, including ports, shipping, and transit systems are also subject to damages, service interruption, and economic losses.

During Ike, much of the most severe impacts on the transportation network were felt in the Galveston area. Damage to County Road 257, also known as the Blue Water Highway, was significant. The portion of roadway from San Luis Pass Bridge to the town of Surfside had to be closed and surge from the storm took out large parts of the road, creating tidal pools where the road had once been. Two miles of roadway were destroyed, 5 miles were

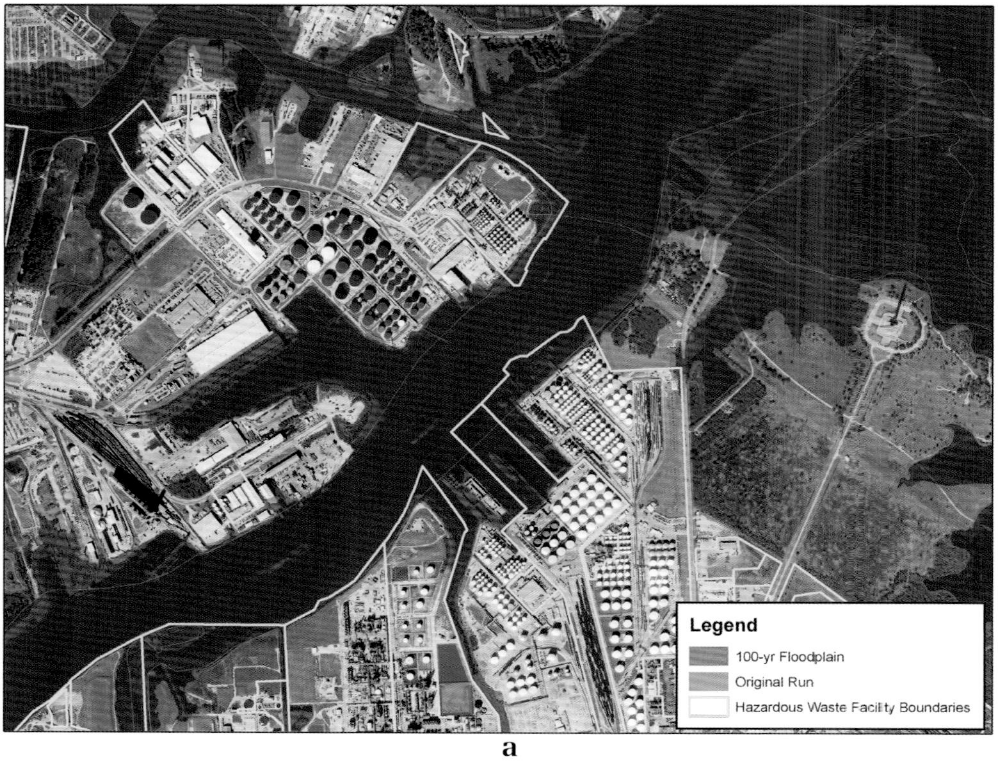

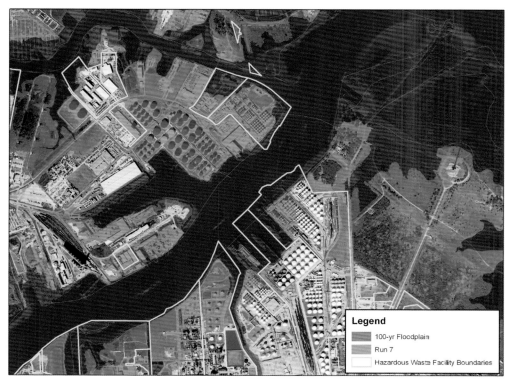

Figure 10.4a. The inundation of hazardous waste facilities in the Houston Ship Channel Area during Hurricane Ike (modeled using ADCIRC).

Figure 10.4b. The inundation of hazardous waste facilities in the Houston Ship Channel Area had winds been 30 percent higher and had the hurricane made landfall near Jamaica Beach (modeled using ADCIRC).

seriously damaged, and erosion was of significant concern. Texas Highway 87, the hurricane evacuation route from Bolivar Peninsula, also suffered extensive damages. In addition, the transit network in Galveston lost half of its bus fleet, para-transit vans, rail trolleys, and rail service.

Damage to the Port of Houston kept it closed for almost a week after the hurricane. Most of the damage was confined to downed trees and power lines, broken glass, and saltwater intrusion in Galveston bayside facilities. The loss of power and failure of the City of Houston lift station that serves the Turning Basin terminals lengthened recovery time. Water breached the docks on the south side of the Turning Basin and at Manchester terminals. The combination of 100 mph winds, waves, and backwash allowed water to rise to within 2 feet of breaching the upstream docks.

The Port of Galveston, a major cargo hub and primary cruise port for the Gulf Coast, was not as fortunate as the Port of Houston. Substantial and widespread damage to structures as well as heavy water damage to equipment, buildings, and piers was evident after Ike. Cruise terminals 1 and 2 were damaged to varying degrees; Terminal 2 sustained heavier damage and was kept closed for more than 4 months. The Port of Galveston was inundated by 8 feet of water at the peak

of the hurricane and almost 1000 vehicles parked at the Port were totaled. Despite all the damages, the Port of Galveston re-opened within a week after Ike.

In addition to port facilities, many piers and dikes were heavily damaged or destroyed. The Texas City Dike, for instance, was overtopped by surge. The Walter Umphrey Pier in Port Arthur, as well as the Galveston fishing pier, the 61st street Pier, and the San Luis Pass Pier in Galveston, were all severely damaged.

**Public and Private Services**

Schools and universities are some of the most important elements of critical infrastructure. They significantly impact large sectors of the population and are indicators of recovery. In addition to University of Texas Medical Branch and Texas A&M Galveston, many campuses of private and public universities and school districts were damaged during Ike. These damages ranged from roof damage to total destruction. Many schools remained closed up to three weeks after the hurricane.

Hurricane impacts on small business cannot be ignored. Small locally-owned businesses are essential to local economies and generally have no means for recovery without assistance. Many small businesses were not expected to return after Hurricane Ike because of damages to their facilities, lost inventory and lost revenue from downtime. It is estimated that in Harris County alone, more than 67,000 business establishments were disrupted by Hurricane Ike (fig. 10.5). More than 1 million workers were unable to work.

Another sector significantly affected by Ike was the health and medical sector. Access to emergency medical services, health clinics, physicians, and hospital emergency rooms was interrupted by Ike. For example, major damage occurred at six hospitals in the greater Houston area, and three in the greater Beaumont area (FEMA 2008). The University of Texas Medical Branch in Galveston sustained damages in excess of $700 million and is the only hospital in Galveston serving a number of surrounding counties. Medical services were unavailable in those communities for a period of time after the event. Hurricane Ike also affected long-term care facilities such as nursing homes and assisted living facilities. Some counties, such as Chambers County, experienced an almost 50 percent decline in their nursing home capacity due to Ike.

**Environmental Impacts**

Environmental impacts from hurricanes are not currently well understood. Releases to the environment may occur in advance of a hurricane as industry prepares for the emergency and secures facilities as well as during the hurricane when susceptible infrastructure fails while encountering the force of wind or water. Pollution in air, soil, and water result from a hurricane due to storm surge, high winds, movement of sediment, and inundation of infrastructure. Thus, there are numerous environmental consequences of an extreme event. For example, after a hurricane, odor pollution is likely to occur from petroleum vapor releases and putrefying sludge, organic matter, or mold. In addition, dust pollution can occur due from deposited sediment that is drying. Superfund sites, toxic waste sites, major industrial facilities, ports, barges, and vessels that handle oil and hazardous chemicals all have the potential to release chemicals and toxins into the environment before, during, and after a hurricane.

Hurricane Ike caused extensive environmental damage. More than 3000 pollution incidents were reported that were caused by the hurricane. Abandoned propane tanks and other hazardous materials washed up on the shore, marshes, and backyards of homes. Approximately a half million

gallons of crude oil were spilled into the Gulf and local bayous and bays. Power failures caused some chemicals to be released into the atmosphere.

Environmental pollution poses a risk to public health in the wake of a hurricane. Those in flooded areas may become ill after contact with floodwater or muck containing high levels of bacteria and other waterborne pathogens from raw sewage. Rashes, blisters, vomiting, and infected sores are not uncommon where skin has come into contact with polluted water. Respiratory problems may occur, as well as potential outbreaks of hepatitis and other waterborne infectious diseases. Petroleum spills and leaks, heavy metals, and other pollutant-laden residue left in the wake of a hurricane are also health hazards.

Figure 10.5. Wind and surge damaged boats at the Houston Yacht club. Photo © 2011 Bryan Carlile, Beck Geodetix.

## Natural Resources

Despite its complexity, it is important to quantify and understand the impact of hurricanes on natural resources. Spills, storm surge, and flood waters can carry salt water and pollution into sensitive waters and marshes that serve as nurseries for birds, fish, shrimp, and other forms of wildlife. Saltwater intrusion in freshwater wetlands destroys these diverse habitats. Often the damage to these ecosystems is overshadowed by the attention paid by the media to the humanitarian aspects of a hurricane event.

Aquatic life is affected in a number of different ways during a hurricane.

Toxic chemicals and wastes, for example, entering coastal waters increase levels of ammonia, dissolved phosphate, and dissolved organic carbon, causing phytoplankton blooms and depleting dissolved oxygen. These conditions, combined with contaminated runoff and petroleum and chemical spills, may result in large fish kills (fig. 10.6). After Hurricane Ike, for instance, thousands of dead redfish were found on the streets of Orange County.

Saltwater intrusion disturbs estuarine habitats, as does the deposition of sediment and debris carried in by hurricane waters. Hurricane Ike had a substantial impact on aquatic resources within the bays and coastal

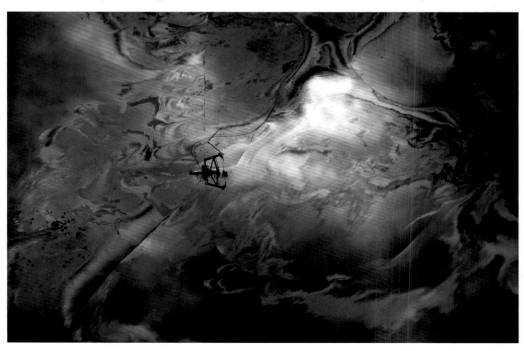

Figure 10.6. Flood waters shimmering with oil on High Island in Texas. Photo courtesy *Houston Chronicle.*

waters. In particular, the affected areas suffered significant damage to their oyster reef habitats and oyster productivity. The oysters in Galveston Bay were covered with a layer of soil, sand, mud, and vegetation brought in by Hurricane Ike, reducing the size of the reefs by 60–80 percent based on side-scan radar mapping that was done by Texas Parks and Wildlife Department. Oysters act as filters for a water system and are primary indicators of system health. In Galveston Bay, the loss of the oyster reefs worsened water quality, weakened biodiversity, and reduced bay productivity. Additionally, oyster reefs provide food and shelter for many species at the base of the food web as well as for speckled trout, redfish, flounder, and croaker. Shrimp and crab populations in Galveston Bay were also damaged, but have since rebounded because Ike flushed the system leaving an abundance of forage on the bay floor due to a mix of soil, sand, mud, and vegetation that was swept from the land into the bay.

Hurricane Ike also affected other wildlife species. A significant number of alligators were found dead along the coastal areas due to saltwater intrusion and the lack of fresh, clean water. In addition, snakes and free-roaming cattle were found in residential and commercial neighborhoods. Many of the cattle were unnaturally aggressive due to distress caused by ingestion of salt water.

The barrier islands around Galveston are some of the best birding locations in the nation. High Island and nearby areas, for instance, are frequented by migrating birds. The hurricane arrived during a heavy migration period to wintering areas; Ike destroyed much of the local food supply preventing birds from feeding for a long time. Many migrant populations survived the storm but the lack of food had a devastating effect on their numbers.

Storm surge and strong winds affect barrier islands that serve as buffers against these natural disasters and are important wildlife habitats. Post-hurricane flights by the US Geological Survey (USGS) showed significant land loss after Hurricane Ike. For instance, the beach ridge in McFaddin National Wildlife Refuge (fig. 10.7) was significantly eroded. Because of its ecological importance, and its role in protecting the salt bayou marsh system from seawater, this has had major impacts on the fresh, intermediate, and brackish marshes in the refuge. Marsh loss has long-term negative impacts on fisheries production and use of the marshes by migrating waterfowl and wading birds.

Erosion also reduces the ability of barrier islands to act as natural buffers to storm surge and flooding. The shape and size of barrier islands change naturally due to longshore currents, wave

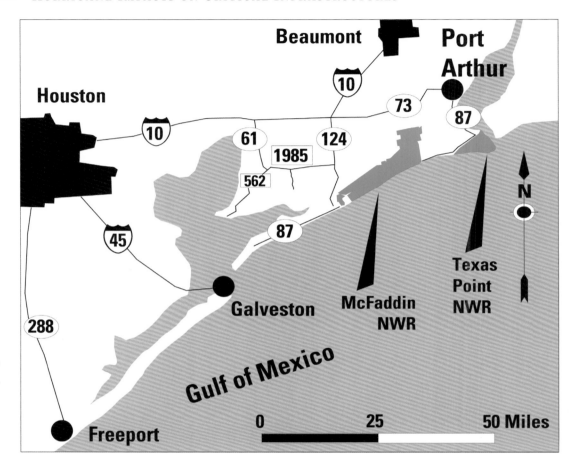

Figure 10.7. Houston/Galveston area ecosystems affected during Hurricane Ike. Courtesy National Wildlife Refuge System.

energy, tides, sea-level change, wind and storms, and the constant movement of sand from their shorelines to their bay sides. Unfortunately, many of these barrier islands have been developed as residential and tourist communities and property damage has become a significant component of hurricane effects on barrier islands. Bolivar Peninsula, a thin barrier beach, is a prime example. The far eastern end of Bolivar Peninsula is known as High Island, named because of the salt dome that is about 38 feet above Gulf Coast typical elevations (fig. 10.8). During Ike, Bolivar Peninsula was partially inundated and High Island was the only area that remained above water.

Hurricanes also impact the ocean floor. Hurricane Ike reshaped the seafloor at the inlet between Galveston Bay and the Gulf of Mexico. The surge from Ike filled Galveston Bay with 12 feet of water, and this surge and back-surge caused erosion and transport of sediment over long distances. Sediment deposition ranged from a few inches to several feet.

## Conclusion

Hurricane Ike ranks as the third most costly disaster in the United States. It caused multi-faceted infrastructure damage and might be most remembered for the extended period of darkness due to power loss. The most significant effects of Ike include destruction of built infrastructure, pollution and damage to ecosystems, shoreline erosion, and disruption of normal function of the region. The lessons learned from Ike emphasize preparedness and the need to develop redundancy in our critical infrastructure.

October 22, 1999 [satellite (Landsat 7)]

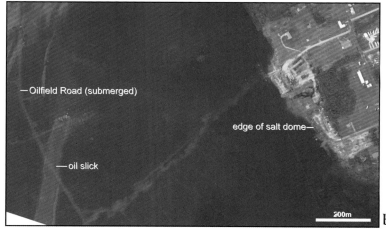

September 16, 2008 (aerial photograph)

Figure 10.8a–b. High Island is a salt dome located on Bolivar Peninsula. During Hurricane Ike, everything but High Island was submerged by storm surge. Photo courtesy NASA.

# 11

# Land-Use Change and Increased Vulnerability

*Samuel David Brody*

## Introduction

With over 50 percent of the US population residing in coastal areas, local decision makers are finding it increasingly difficult to protect critical natural resources, and facilitate the development of hazard-resilient communities. Nowhere is this more apparent than on the Texas coast. Rapid urban and suburban development has resulted in loss of critical habitats and key species while at the same time placing human populations in areas vulnerable to natural hazards. These problems are exacerbated within major population centers, particularly the Houston/Galveston area, where population growth, sprawling development patterns, and the alteration of hydrological systems have created some of the most vulnerable communities in the nation.

The following sections trace the causes and consequences of development within coastal watersheds with special emphasis on flooding in Texas. The underlying premise is that human exposure to natural hazards, such as floods and hurricanes, is not solely a technical or engineering problem, but one driven by land-use change and the pattern of development across metropolitan regions. First, the causes of land-use change within coastal landscapes are addressed based on the following four factors: population growth, spread of impervious surfaces, loss of naturally occurring wetlands, and sprawling patterns of development. Next, the adverse consequences of land-use change with respect to flood damage, social vulnerability, and risk exposure to severe storms are addressed. Finally, the policy and planning implications for mitigating the impact of the built environment and more effectively protecting communities from the threat of coastal hazards in the future are discussed.

## Background and Problem Statement

Given the recreational, aesthetic, and economic opportunities available on the coast, this geographic area has historically been the focus for exten-

sive population growth and land use change. In 2003 for example, it was estimated that approximately 153 million people (53 percent of the population of the nation) live in the 673 US coastal counties, an increase of 33 million people since 1980 (Crossett et al. 2004). As coastal population has increased, more structures have been placed in areas susceptible to the adverse impacts of the severe storms that routinely strike coastal areas. From 1999–2003 for example, 2.8 million building permits were issued for the construction of single-family housing units (43 percent of the total for the nation) and 1 million building permits were issued for the construction of multi-family housing units (51 percent of the total for the nation) within coastal counties across the United States (Crossett et al. 2004). Because communities positioned along a coastline or within a coastal watershed are especially vulnerable to surge, flooding, and high winds, this upward trajectory of growth has created the ideal conditions for human disasters (fig. 11.1).

Figure 11.1. When Hurricane Ike struck, homes on 8th Street near Galveston were severely damaged. Photo courtesy Greg Henshal / FEMA.

In particular, flooding and resulting flood damage within coastal areas has had a major impact on the national economy. Flooding is the most ubiquitous and costly natural hazard in the United States, causing billions of dollars in property damages each year. Using the National Weather Service (NWS) *Storm Data* publications, Mileti (1999) estimated property losses from floods to be from $19.6 billion to $196 billion between 1975 and 1994 alone. While damage estimates from floods vary, the economic costs from floods appear to be steadily increasing (Pielke and Downton, 2000). For example, Birkland et al. (2003) reported flood damages from 1900–20 totaled $1.76 billion compared to $4.4 billion from 1980–2000. According to data extracted from the *Spatial Hazard Events and Losses Database for the United States* (SHELDUS), the average annual flood count has increased sixfold from 394 floods per year in the 1960s to 2444 flood events a year in the 1990s. SHELDUS data also show increasing property damage from floods over time. In the 1960s, floods caused $45.65 million dollars in damage per year; by the 1990s, average annual property damage from flooding increased to $19.13 billion dollars a year (inflation adjusted at 1960 dollars) (Brody et al. 2007a). These damage estimates help confirm what has been understood by local decision-makers for over a decade: that floods and severe storms are a major risk to the health and safety of the US population and with increasing development in low-lying coastal areas, the problem appears to be getting worse.

Hurricane Ike illustrates the negative impacts of such events where large populations live in physically vulnerable areas. Preliminary damage estimates by FEMA include $3.4 billion for housing, with approximately 8000 housing units destroyed over several counties. A recent damage assessment conducted by Texas A&M University for Galveston Island, where Ike made landfall, showed that 10 percent of all structures were completely destroyed (fig. 11.2). Several months after the hurricane, 40 percent of single family houses and 75 percent of multi-family units remained vacant. The State of Texas estimates that repairs to waterways and ports will cost more than $2.4 billion. Fixing water and wastewater plants and government buildings will take at least $1.7 billion. In addition, it is expected that more than $1 billion will be needed to repair schools and universities.

**Characteristics of Coastal Land-Use Change**

In his book *Disasters By Design*, Mileti (1999) pronounced that disas-

ters do not simply happen as "acts of God." but are largely the result of how we build and design human communities. Building on his assertion, this chapter emphasizes that the location, intensity, and pattern of the built environment are the critical factors in determining the impact of a severe storm. The section below identifies and discusses four main characteristics of land-use change associated with the built environment in Texas.

### Population growth

The first characteristic of land-use change and resulting exposure to flood risk is placing increasingly more people in physically vulnerable areas. Between 1980 and 2003, there were approximately 2.5 million people added to coastal areas in Texas, representing a 53 percent increase (fig. 11.3a, b). Only California and Florida ranked higher (Crossett et al. 2004). Surprisingly,

Figure 11.2. Storm surge exposed geotubes that were acting as semi-natural sand dunes and homes were destroyed. Photo © 2011 Bryan Carlile, Beck Geodetix

the Texas coast remains relatively un-developed, with population centered in the Houston-Galveston area, Corpus Christi, Beaumont, and Brownsville. However, these populated areas have recently become places of intense growth. For example, from 1980–2003 the coastal growth rate of Harris County, where Houston is located, was the second highest among all counties in the United States (Crossett et al. 2004). Today, the Houston-Galveston-Brazoria region remains one of the fastest growing areas in the country, with over 2100 persons per square mile. The more people that reside in this low-lying, vulnerable coastal area, the more difficult it will become to evacuate communities in advance of a storm and the more likely property and lives will be lost. Recent storms, such as hurricanes Rita and Ike have demonstrated these difficulties. Given the expected increases in population along the Texas coastal plain, the scale of these disasters will only become amplified in the future.

*Increased area of impervious surfaces*

Increasing numbers of people living, working, and playing along the coast generates more impervious surfaces. Thus, the rising intensity of land-use change does not simply translate into more people in harm's way, but also an increase in hardened surfaces that compromise existing hydrological systems and their drainage regimes. The Houston-Galveston area has quickly become one of the largest expanses of impervious surfaces in the country, with a literally uninterrupted swath of pavement approximately 60 miles long and 40 miles wide (fig. 11.4). Freeways, parking lots, rooftops, and urban parklands are ubiquitous across the Houston metropolitan area. If water is unable to drain slowly into the soil or nearby water bodies, it has nowhere to go but into homes and businesses. For example, in Houston a routine summer rain shower of 4 inches is enough to flood major roadways. Water pools onto road surfaces (particularly highway underpasses), which are usually the lowest-lying areas in the city, trapping motorists every year.

*Loss of naturally occurring wetlands*

Rapid land-use change and resulting impervious surfaces alone increase the risk of flooding and associated flood damage during severe storms. However, this risk is significantly amplified when pavement and other hardened surfaces take the place of naturally-occurring wetlands. Thus, the specific location of development within the hydrological landscape becomes an important predictor of increased flood risks. In fact, numerous hydrological studies demonstrate the effectiveness

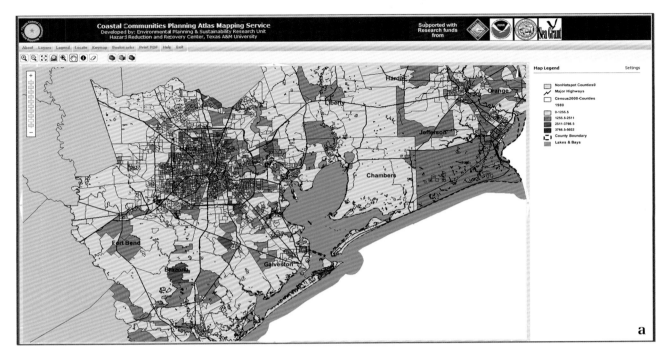

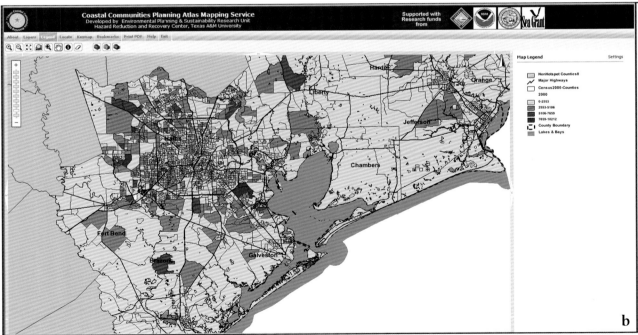

Figure 11.3a–b. These figures show the difference in population density for the Houston/Galveston area between 1980 (a) and 2000 (b).

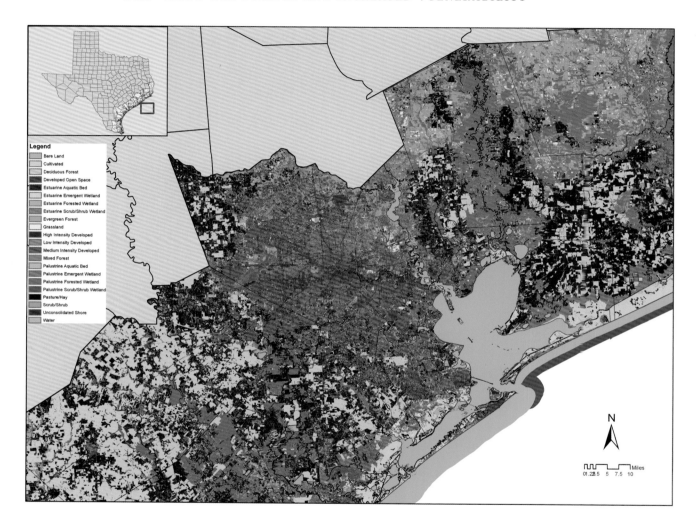

Figure 11.4. Land use in the Houston/Galveston area.

of coastal wetlands to attenuate the intensity of flood events (Mitch and Gosselink 2000; Bullock and Acreman 2003) and mitigate resulting property loss and human casualties (Brody et al. 2007b; Zahran et al. 2008).

Texas ranks among the top five states in the country in terms of total wetland area with an estimated 3 million hectares, comprised largely of palustrine and estuarine wetlands (Dahl 2000). However, the record of wetland alteration along the Texas coast over the last several decades suggests that the hydrological systems are being fragmented and the ability of these systems to reduce flooding during severe storms is being compromised.

A spatial analysis of wetland alteration permits under Section 404 of the Clean Water Act, from 1991 to 2003, demonstrates the intensity and pattern of land use conversion associated with this critical natural resource (Brody et al. 2008) (fig. 11.5). In coastal Texas, the majority of wetland permits during this 13-year time period impacted estuarine systems. Estuarine or tidal fringe wetlands in Texas are mostly found between the open salt water of the bays or Gulf and the uplands of the coastal plain and barrier islands. Loss of estuarine wetlands coincides with the concentrated development patterns adjacent to coastal waters, particularly around Galveston and Corpus Christi bays. Palustrine wetlands also comprise a significant percentage (almost 36 percent) of wetland alteration permits. These development activities are most likely to take place farther inland and off the direct coastline in non-tidal or tidal areas where salinity due to ocean-derived salts is below 0.5 parts per thousand.

In coastal Texas, wetland alteration occurs over a surprisingly large area in and around the Houston metropolitan area where palustrine wetlands have been heavily impacted. Even though the heaviest growth in the region is yet to come, the study tallied over 857 permits granted per year within the coastal zone. A relatively large number of "general" permits were also granted in Texas. This type of permit is most often associated with oil and gas production activities pervasive in parts of eastern Texas. A general permit category may include providing industry with the rapid authorization needed to construct pipelines, wells, and other oil and gas activities (Brody et al. 2008). In general, both the intensity and spatial pattern of wetland alteration via the federal permitting process are diminishing the function of hydrological systems to store and slowly release runoff.

*Outwardly sprawling development patterns*

Coastal land-use change contributing to increased vulnerability to severe storm events is not only a function of the intensity and location of development, but also its regional pattern. Sprawling development patterns, typified by low-density, residential dwelling units spreading outward from urban cores, dominate the Texas landscape. The result of this kind of development pattern is the over-consumption of land originally designated for other purposes. For example from 1982–97, over 1.7 million acres of agricultural land were developed, more than any other state in the nation (USDA 2000). The Houston-Galveston area is perhaps the best example of urban and suburban sprawl in Texas. From 1970–90, approximately 640 square

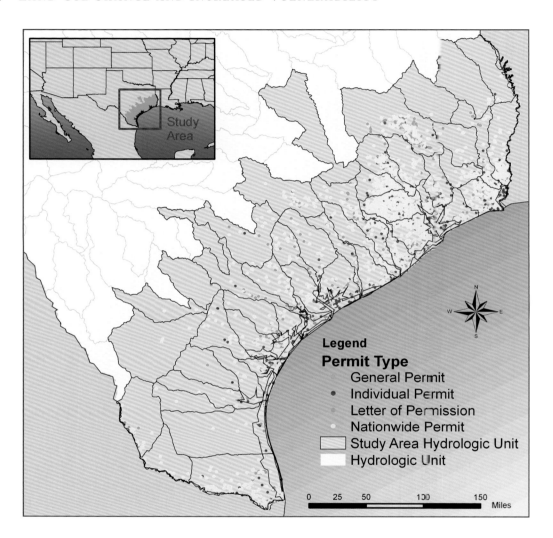

Figure 11.5. Wetland alteration permits along the Texas coast.

miles of land in the Houston area was urbanized, second only to Atlanta, Georgia during the same time period (US Census Bureau on Urbanized Areas, as cited on www.sprawlcity.org).

Another indicator of sprawling development patterns is the spatial record of federal wetland alteration permits described above. Additional analyses of the data show that from 1991–2003, 78 percent of all wetland permits were issued in an increasingly dispersed pattern outside of urban areas. Furthermore, over 72 percent of these permits were categorized by the US Army Corps of Engineers (USACE) as "nationwide," which usually entails small-scale residential development activities having minimal individual adverse impacts on the aquatic environment (Brody et al. 2008). Nationwide permits, used to alter primarily palustrine wetlands, are ubiquitous in the Houston-Galveston region, suggesting that these wetland systems are being cumulatively impacted from the smaller-scale developments symptomatic of sprawl. Sprawling patterns of residential development in coastal Texas are further exacerbated because: a) there are no large protected areas to buffer outward growth as is the case in other states such as Florida; and b) there are no mandated growth management or comprehensive planning regulations that could help focus development in urban areas. The result of this spatial trend in the built environment is the conversion of agricultural lands, increased area of impervious surfaces, loss of naturally-occurring wetlands, and increased vulnerability to the adverse effects of severe coastal storms.

## Consequences of Changing Coastal Landscapes

One of the consequences of rapid coastal land-use change in Texas is increased flooding and flood damage. Texas consistently leads the nation in property damage as a direct result of flooding. According to Federal Emergency Management Agency (FEMA) statistics on flood insurance payments from 1978 to 2001, Texas experienced $2.25 billion dollars in property loss, more than California, New York, and Florida combined (http://www.fema.gov/nfip/pcstat.htm). Not only are floods frequent in Texas, but like everything else in the state, they tend to be large. For example, of the 42 flood events listed as causing more than a billion dollars in damage from 1980–98, four were in Texas (NCDC 2000). In a recent study of 37 coastal counties, Brody et al. (2007b) found that 423 flood events caused over $320 million in reported property damage from 1997–2001. The average amount of damage per flood in the study area, which extended along the entire Texas coastline, was $423,765.90. The most deaths in the nation from flooding are also consistently reported from Texas, and average twice those in California, the state with the next highest deaths (Frech 2005). In a parallel study examining 99 counties and 832 flood events in eastern Texas from 1997 to 2001, Zahran et al. (2008) catalogued nearly 6000 deaths and injuries, an outlier when considering these figures relative to other states.

The startlingly high degree of property loss and human casualties from flooding in Texas can be traced to the intensity, location, and overall pattern of development across coastal landscapes. For example, due to the increase in impervious surfaces, urbanization is a primary contributor to flood frequency (Schuster et al. 2005). Conversion of agricultural and forest lands to urban areas can diminish the ability of a hydrological system to store and slowly release water, resulting in increased flood intensity (Carter 1961; Tourbier and Westmacott 1981). As the area of impervious surface coverage increases, there is a corresponding decrease in infiltration and an increase in surface runoff (Dunne and Leopold 1978; Paul and Meyer 2001). According to Arnold and Gibbons (1996), as the percentage of impervious surface within a drainage basin increases to 10–20 percent, corresponding runoff doubles. Greater surface runoff volume often results in increased frequency and severity of flooding in streams. Brody et al. (2007c) found that an increase in impervious surfaces (as measured through remote sensing imagery) across 85 coastal watersheds in Texas and Florida was positively correlated with a significant increase in stream flow over a 12-year period. In coastal Texas from 1997–2001, every square meter of added impervious surface translated into approximately $3,602 of added flood damage per year (Brody et al. 2007b).

Impervious surfaces alone increase the risk of flooding and resulting flood damage, but the problem is exacerbated when it replaces naturally occurring wetlands. When controlling for multiple socioeconomic and geophysical contextual characteristics, Brody et al. (2007b) found that the loss of wetlands was the most powerful predictor of reported flood damage among all the variables measured for the built environment in coastal Texas. In fact, each permit to alter a naturally-occurring wetland granted by the USACE under Section 404 of the Clean Water Act translated into an average of $211.88 in added property damage per flood over the 5-year period from 1997–2001. While this economic impact may seem miniscule, considering there are tens of thousands of permits granted across coastal Texas, the cumulative impact of wetland loss becomes a prominent factor in triggering flooding. For example, on average, wetland alteration adds over $33,000 in property damage per flood. Even more revealing, only 129 Section 404 permits equal the flood-reducing effects of a dam in coastal Texas over the study period. While dams and other structural engineering flood mitigation can cost over $100 million, protecting naturally-occurring wetlands is relatively inexpensive.

As suggested above, it is important to note the precise location of development associated with coastal land-use change. Not all wetlands are equal in terms of their contributions to flood attenuation. For example, wetland alteration inside the 100-year floodplain and in close proximity to main-stem rivers and other water bodies, significantly raises the probability of flooding in urban and suburban areas. Of the approximately 8 million structures in Texas floodplains, 3 million are uninsured (Frech 2005). Brody et al. (2008) calculated that nearly 40 percent of all wetland development sites in coastal Texas were located within the 100-year floodplain. This percentage actually increases to almost 60 percent in the Houston-Galveston area. The scale of alteration is also important. A project that requires filling or replacing more than half an acre of wetland has a significantly larger negative effect than smaller developments. However, because there are so many smaller-scale projects, the cumulative impacts from nationwide permits cannot be overlooked. The "death by a thousand cuts" phenomenon of micro-scale adjustments to the function of hydrological systems may be the greatest challenge for local decision makers interested in facilitating the development of flood-resilient coastal communities.

The overarching pattern of the built environment is another characteristic of land-use change that is exacerbating flood problems in coastal Texas. Outwardly sprawling, low-density residential development, particularly in and around Houston, not only leads to increased impervious surfaces and wetland alteration, but also exposes more people and structures to risk from severe storms. As rooftops spread outward in ever-expanding bands to the northwest and southeast of Houston, the community-level exposure to risk associated with flooding is greatly increased. In fact, the change in risk exposure from 1980–2000 practically mimics the shift in population during the same time period in this area (fig. 11.6). Sprawling development that encroaches into agricultural lands makes it more difficult to provide critical facilities and/or evacuate ahead of an approaching storm compared to more concentrated urban areas. Examining over 4000 US Census block groups along the entire Texas coast over two decades, further accentuates this problem. From 1980 to 2000, for example, 63 percent of all block groups increased in both population and exposure to meteorologically-driven risks. By comparison, only 3 percent of these blocks increased their population, but decreased their risk (fig. 11.7).

Figure 11.6. Change in population exposed to meteorologically-based risk, 1980–2000.

Figure 11.7. Harris County floodplains identified by FEMA. Map courtesy Harris County Public Infrastructure.

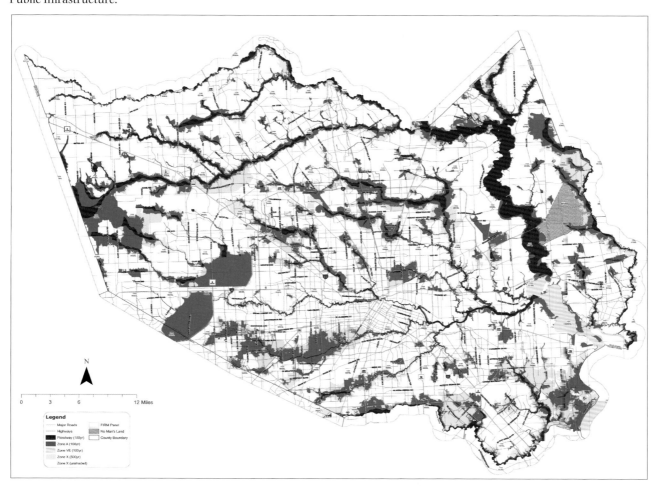

## National Flood Insurance Program (NFIP)

In addition to economic and aesthetic considerations, settlement along the coast and in or near flood plains may be partially attributed to the unintended effects of the National Flood Insurance Program (NFIP). Established in 1968 in response to a lack of affordable flood insurance to home owners, the purpose of the NFIP was two-fold: 1) to help provide financial assistance for victims of flood disasters to rehabilitate their property; and 2) to help prevent unwise use of land where flood damages would mount steadily and rapidly. By requiring that homeowners within the flood plain purchase insurance, the NFIP would be able to fund itself using premium payments while also providing an economic disincentive for settling in a high risk area.

Initially, the NFIP attempted to manage its risk by requiring communities to take part in enumerated mitigation strategies before becoming eligible for flood insurance. However, after five years only 200,000 policies were held because many communities did not view insurance as a strong enough incentive to undertake the required mitigation strategies. Enforcement also proved difficult and the program was in danger of dissolution. To prevent this, Congress passed the Flood Insurance Protection Act of 1973, mandating that federal emergency disaster relief would only be provided to those communities that participated in the NFIP. Membership grew as a result of the Flood Insurance Protection Act resulting in over 19,000 participating communities and over 4.4 million policies by 2002.

In addition to the Flood Insurance Protection Act of 1973, membership also grew as a result of very affordable premium rates; so affordable in fact that the rates began to inaccurately reflect the actual cost of residing within the flood plain. This inaccuracy initiated a tipping point for settlement along the coast. While the initial goal of the NFIP was to provide economic assistance to those already living within the flood plain while discouraging over-use, the NFIP now provided an economic incentive for settling within the floodplain. Once this switch occurred, the NFIP was no longer able to fund itself using profits from premium payments resulting in annual operating losses in the years 1972–80, 1983, 1984, 1989, 1990, 1992, 1993, 1995, and 1996 (Burby 2002).

In addition to the issue of premium rates, the NFIP also suffers from lack of accurate information. Because premium rates are a reflection of the location of a structure relative to the 100-year floodplain, accurate flood insurance rate maps are essential if accurate rates are going to be assigned.

However, due to the large cost and time elements associated with mapping, only 12,000 of the more than 19,000 communities participating in the NFIP possess flood insurance rate maps. In addition, the maps that do exist may not accurately represent the 100-year floodplain. To underscore this idea, from 1978–96, 23 percent of the NFIP claim dollars were for communities that existed outside the 100-year floodplain. Without accurate maps, it is difficult to verify that risk is evenly distributed among insured communities. This issue of risk coupled with an economic incentive to settle within the 100-year floodplain has led to a 53 percent increase in development in the floodplain over 30 years, (Burby 2002) thus introducing increased risk into an already unbalanced system.

**Policy Implications**

Most population projections place the Texas coast among the fastest growing regions in the nation over the next several decades. The large amount of relatively undeveloped land coupled with shifting urban and suburban populations will exert development pressure on ecologically-sensitive and vulnerable coastal systems. The risks and damages from storms will continue to mount unless we alter the way in which we design and develop the land-

scape. The following policy recommendations can help mitigate adverse consequences from storms as more people live, work, and play along the Texas coast.

*Avoid the floodplain*

One of the most basic design elements for building resilient communities is to stay out of the 100-year floodplain (where there is a 1 percent chance of flooding each year). Time and time again, we facilitate the development of houses, schools, and critical facilities in flood-prone areas. Over the course of a 30-year mortgage, there is a 26 percent chance that a home in the floodplain will experience flooding (NRC 2000). In coastal areas, the floodplain is generally well-documented and should serve as a planning guide for community development. Filling areas of the floodplain, particularly naturally-occurring wetlands, or raising sites above flood elevation may serve as a protective measure for a specific development, but flooding risks are often increased downstream. The cumulative impacts of seemingly minor alterations can compromise the ability of hydrologic systems to store runoff and alter the boundary of a floodplain. In several instances, older neighborhoods in the Houston area that are not historically in the 100-year floodplain started to experience flooding due to more recent development in surrounding areas. In these cases, development inside the floodplain expanded the original boundaries of the floodplain outward and placed uninsured homes at risk to inundation.

*Consider wetlands for flood control*

From a policy-making perspective, the United States is well-beyond the notion that wetlands are unusable swamps that should be eliminated for the good of human settlements. We now protect and celebrate their multiple values including water quality, wildlife habitat, commercial harvesting, and places for recreation and tourism. However, rarely do decision-makers at any jurisdictional level systematically consider naturally-occurring wetlands as effective flood control devices. There is convincing evidence, some of which is presented above, that maintaining these areas is the most cost-effective and easiest way to protect communities against flooding associated with coastal storms, even compared to dams and other structural engineering techniques.

If it is necessary to alter a naturally-occurring wetland, the loss of flood protection should at least be factored into the development. Now that we can put a price on the alteration of wetlands in terms of flood damages and other impacts (Highfield and Brody 2006), it seems reasonable that these costs should be internalized in the development process. Permitting fees, impact fees, and public infrastructure costs should all reflect the economic burden these developments place on the larger community. For example, several localities in Florida exact a wetland alteration permit fee in addition to existing state and federal permitting requirements to account for the values lost to the community when a naturally-occurring wetland is lost to development.

*Adopt nonstructural flood mitigation techniques*

Across the country, localities are increasingly embracing both regulatory and incentive-based nonstructural flood mitigation techniques as effective alternatives to traditional structural engineering projects. These strategies include a range of options such as land-use planning tools, tax incentives, education and training, environmentally sensitive area protection, flood forecasting models, and other emergency and recovery policies for mitigating flood loss (Whipple 1998; Birkland et al. 2003). Nonstructural mitigation techniques can be adopted by communities through several mechanisms, such as comprehensive plans, stand-alone watershed plans, and local mitigation strategies.

One relatively new venue for nonstructural flood mitigation is FEMA's Community Rating System (CRS). Adopted in the early 1990s, this program encourages communities to go beyond the minimum standards of the NFIP for floodplain management by providing discounts of up to 45 percent on flood insurance premiums for residents of participating communities. Credit points are assigned for 18 activities organized into the following four broad categories of floodplain planning and management: public information, mapping and regulation, flood damage reduction, and flood preparedness. Premium discounts correspond to points accrued by each participating community. Discounts range from 5 (class 9) to 45 percent (class 1) depending on the degree to which a community plans for the adverse impacts of floods. Recent empirical evidence suggests that the CRS is effective in reducing flood losses. In the coastal counties in Texas from 1997 to 2001, Brody et al. (2007b) found a real unit increase in CRS class equaled a $38,989 reduction in average property damage per flood. If every jurisdiction in the study area had maximized their CRS rating, the cost of floods would have been less than a quarter of $320 million observed.

Despite the promise of nonstructural mitigation and planning techniques, they are used rather sparingly in coastal communities in Texas. A recent survey of coastal localities found only 36 percent made extensive use of stand-alone flood mitigation plans. Furthermore, approximately 45 percent of the communities sampled had never used development setbacks, over 60 percent never used protected areas, and almost 55 percent never used land acquisition as strategies to mitigation the adverse impacts of floods (Brody et al. 2009). Communities in Florida, by comparison, reported significantly more use for nearly all fourteen nonstructural flood mitigation activities evaluated in the study.

*Adopt sprawl reduction planning policies*

An important subset of nonstructural mitigation policies can be categorized specifically for the reduction of sprawl. As noted above, outwardly expanding, low-density suburban communities are placing more people and associated structures in areas vulnerable to severe storms. Planning tools that contain and focus development within central business districts and away from ecologically sensitive areas can help reduce the adverse impacts of floods. Specific land-use strategies, such as transfer of development rights (TDR), development density bonuses, clustering of houses, and conservation easements, to name a few, can together foster more resilient and sustainable local communities over the long term. These are commonly-used policies in many parts of the country because of their demonstrated effectiveness (Brody et al. 2006). If decision-makers in Texas want to stem the rising costs of floods, adopting sprawl reduction policies may be a necessity.

**Conclusion**

Coastal Texas is a national leader when it comes to property losses and human casualties from floods in almost any given year. Addressing increasing losses from flooding as a result of both chronic and severe storms cannot be accomplished through solely structural and engineering solution to mitigation. Decision-makers, developers,

and residents must also consider the location, intensity, and overall pattern of the built environment. As history has shown, the key to building flood resilient communities is not in erecting dikes and dams, but in the regional urban and suburban form we choose to establish across coastal landscapes. Thus, a future agenda for research should be constructed around understanding which features built environment enabling development without damage. The results of these inquiries will inform practitioners as to which set of policies they should adopt to most effectively protect the economic interests and overall livelihood of localities over the long term.

# 12

## Steps to the Future

### Jim Blackburn, Thomas Colbert, and Kevin Shanley

**Introduction**

Hurricane Ike was not a worst case storm for our region. Ike came ashore to the east of metropolitan Houston and caused most of its damage through surge flooding to Galveston Island, the Bolivar Peninsula, and areas immediately adjacent to Galveston Bay, with more widespread wind damage over the Houston region. Major surge damage also occurred eastward into the Sabine Lake watershed. Although Ike caused upwards of $24 billion in damage, the damage could easily have reached $100 billion if it had come ashore further south, in the San Luis Pass area. More than anything, Ike exposed the vulnerability of the Houston-Galveston region to a major storm.

In the aftermath of Ike, a call for responsive action was sounded, first by Dr. William Merrell in his proposal for the "Ike Dike," a levee and sea gate structure extending from Freeport to High Island, and then in the report of the Governor's Commission for Disas-

ter Recovery and Renewal. Following this report, the Gulf Coast Community Protection and Recovery District was created to study and potentially finance infrastructure improvements in Orange, Jefferson, Chambers, Galveston, Harris and Brazoria counties.

The key question facing the region is what should be done to prepare and protect the region in the event that another storm such as Ike or one with higher winds and a more substantial surge tide hits the region. In the sections that follow, a series of alternative approaches to become more resilient are set out. However, prior to discussing various alternatives, it is useful to consider certain aspects of the regulatory and funding climate that are emerging in late 2010.

**Challenges in Obtaining Federal Funding**

As will be discussed in later sections, the size and costs of certain structural and even nonstructural alternatives are large. Although money to initiate detailed studies of these alternatives may be generated locally, it is very unlikely that the money to implement one or more of these alternatives will come from either state or local government sources. Instead, the Houston-Galveston region will likely look to the federal government to obtain funding, relying upon the US Army Corps of Engineers (USACE) and the United States Congress to provide money. In the past, the region has benefitted significantly from federal largesse along the coast; however, the rules and realities of 2010 are different from those in effect when levees were built around Texas City and Freeport. A short survey of these rules and changes is useful in helping formulate the next steps needed for regional action.

## National Environmental Policy Act

First, the applicability of the National Environmental Policy Act (NEPA) must be considered. NEPA was passed in 1969 and has long been applicable to major federal actions. NEPA is a procedural act that requires the full development and articulation of alternatives and their environmental impacts in an environmental impact statement. The rules implementing NEPA are clear in this regard. The development and comparative evaluation of alternatives is the "heart" of the environmental impact statement process. The NEPA process requires that "no action" be evaluated along with a full range of "fundamentally different ways of achieving project purposes." It requires honest and transparent development and evaluation of alternative courses of action and provides a mechanism for review of such actions by the federal court system.

NEPA is procedural rather than substantive. However, if there are substantive rules and regulations that interact with the NEPA alternatives and disclosure requirements, NEPA becomes a force that will shape federal decision-making and ultimately alternative selection. In addition to older statutes such as the Endangered Species Act, there are several new developments that will shape project development in the future.

## Emerging Sustainability Concepts

If federal money is to be obtained for a flood surge control project, the likelihood is that the USACE will be the lead agency. In turn, the US Army Corps of Engineers (USACE) evaluates projects under criteria developed according to US Water Resources Council in the "Economic and Environmental Principles and Guidelines for Water and Related Land Resources Implementation

Studies" first drafted in 1983 to guide federal agencies addressing water resource development (Watt 1983). However, the principles and guidelines have been in the process of being changed by the USACE since the Water Resources and Development Act of 2007. More recently, President Obama instructed the Council on Environmental Quality (CEQ) to assist in rewriting of the principles and guidelines as set out in 74 Fed. Reg. 31415, July 1, 2009.

The proposed changes are significant. Most importantly, the CEQ and the USACE would alter project approval criteria to emphasize evaluation and satisfaction of criteria relative to economic, environmental, and social factors rather than just with regard to economic benefits. In other words, the proposed revisions require equal consideration of social and environmental concerns with economic concerns. One objective is to protect and restore natural ecosystems and the environment while encouraging sustainable economic development. Another objective calls for avoiding the unwise use of floodplains, flood-prone areas, and other ecologically valuable areas. Yet another criterion provides that studies shall give "full and equal treatment to nonstructural approaches."

These proposed changes to the US Water Resources Council principles and guidelines are followed in other presidential documents that have been issued since early 2009. Among these changes is Executive Order 13547 issued July 19, 2010, that emphasizes stewardship of the oceans and coasts as well as environmental sustainability. Similarly, the Gulf Coast Ecosystem Workgroup released its *Roadmap for Restoring Ecosystem Resiliency and Sustainability* on March 4, 2010 and found that " . . . the Federal Government and States must work in partnership to recast river and coastal management priorities so that ecosystem restoration and sustainability are considered on a more equal footing with other priorities such as manmade navigation and structural approaches to flood protection and storm risk reduction." Additionally, following the BP Deepwater Horizon oil spill, Secretary of the Navy Ray Mabus has been tasked with coordinating the multiple interest groups and governments to formulate a long-term Gulf Coast restoration plan.

From the foregoing, it is clear that two important trends are emerging. First, it is clear that structural and nonstructural alternatives are to be considered on an equal footing. Second, environmental, economic, and social criteria are to be integrated in the project selection process. In essence, the federal government is adopting the concept of sustainable development, often called the "three-legged stool" or the "triple bottom line" because it requires the integration of economic, environmental, and social issues in project decision-making. It is sustainability that will emerge as the key criteria to the identification of a successful federal project.

## Federal Budgetary Challenges

In addition to the more general challenge of sustainability, there will be substantial challenges to project funding provided by the current status of the US economy and its debt structure. Today, the US government is more heavily in debt than at any time since World War II. In 2009, the national budget deficit was 20 percent of the nation's gross domestic product (GDP) in that same year. The ratio of the nation's debt to GDP is over 50 percent for the first time since World War II. Additionally, borrowing at the global level is at unprecedented levels, with global debt rising from $18 trillion in 2001 to a projected $45 trillion in 2011. Never before have so many countries been so far in debt.

In response to this fiscal situation, a study was conducted by the non-partisan Congressional Budget Office (CBO) entitled *The Budget and Economic Outlook: An Update, August 2010.* According to this document, a fiscal crisis is likely if the growing federal debt is not resolved. In this document, the CBO concludes that "putting the nation on a sustainable fiscal course will require policymakers to restrain the growth of spending substantially, raise revenues significantly above their average percentage of GDP of the past 40 years, or adopt some combination of those approaches."

The bottom line is that for the next several years, the availability of extensive federal funding for major structural flood control projects will be limited to the extent that they are available at all. At the same time, nonstructural alternatives, which are often cheaper, will emerge on an equal footing with more traditional structural alternatives and environmental and social factors will be added to classic benefit-to-cost analysis. If these trends are maintained, they will substantially alter the dynamics and ultimately the likely choice of alternatives that will be federally funded.

## Philosophy of the Alternative

Regional thinking about hurricane issues should initially be focused upon developing as many diverse and reasonable alternatives as can be conceived. Alternative development is a key aspect of creative and sustainable design. In fact, the philosophy of the alternative is a key component of sustainable design as it is taught at Rice

University. Inherent in this philosophical approach is the realization that preconceptions and past practices often keep us from asking important questions and pursuing better ways of proceeding into the future. In this context, alternative analysis includes different ways of meeting desired goals as well as the placement of differing solutions at different locations.

A benefit of thinking in terms of alternatives is that this way of thinking is compatible with the process required for compliance with federal environmental laws such as NEPA as well as the proposed revision of the US Water Resources Council's principles and guidelines. The key to alternatives analysis is to develop fundamentally different ways to address the hurricane-related challenges that face our region. No alternative should be off the table initially, including doing nothing. In fact, a thorough understanding of the impact of doing nothing (e.g., our current vulnerability) should be the starting point, because, among other things, it is required by NEPA for any federal project and is also a likely outcome because it is hard to gain momentum for change.

Alternatives can be divided into structural and nonstructural solutions as well as into shorter-term and longer-term solutions. In fact, the timing of the implementation of alternatives may be one of the more im-

portant concepts to keep in mind. A large-scale structural alternative will likely take much longer to implement than a non-structural alternative or a smaller-scale structural alternative. At the very least, strategies should be conceived considering both the short and long term. Often, the pursuit of larger-scale alternative results in nothing being done for a decade or two while the region waits for federal assistance, only to have the implementation of the alternative delayed by failed federal funding or other mishaps.

It is also important to note that many of the cheaper and easier alternatives to implement are also controversial because they may run counter to certain prevailing belief structures. For example, at the local governmental level land-use regulations may be very effective in addressing hurricane flood damage prevention but these alternatives may be difficult for some members of the community to accept. On the other hand, to put aside these alternatives without considering them is to deny important choices to the community.

Sustainable development requires a full articulation of alternatives including the economic, ecological, and social costs and benefits of the various alternatives. The importance of the development of alternatives is that these costs and benefits can be compared. We need an honest evaluation of these alternatives. We need help in making good decisions. There are many excellent thinkers with preconceptions about choices. Frankly, we need high quality information about multiple alternatives without the burden of preconceptions so that we can make the best decisions for the region in both the short and long term.

## Structural and Nonstructural Alternatives

The near-miss with Ike has led to substantial concern about the potential for significant damage and need for addressing the storm surge problem. There are various ways to address flooding. In the broadest sense, there are structural and non-structural alternatives.

Structural alternatives are familiar to us. These include dikes and levees that keep floodwaters out of certain areas. Levee systems have been used along the Texas coast to protect from coastal flooding, with levee systems existing at Port Arthur, Texas City, and Freeport on the upper Texas coast. There are also structural alternatives that involve "hardening" of individual structures by increasing the strength of the constructed building, raising slab elevations, and other actions intended to make structures able to withstand the ravages of a storm.

There are also other structural alternatives that have been used for stream flooding, such as flood storage reservoirs and channelization, that are probably not applicable to hurricane surge flooding.

Nonstructural alternatives are quite different than structural ones. The basic concept of nonstructural solutions is that we attempt to avoid the conflict between the storm surge and land development by understanding the surge and working with the land development process. As a general proposition, this means directing land development to safer areas while recognizing and protecting the natural function of the low-lying coastal lands and flood plains. Here, there are at least two different types of situations that are encountered—addressing the conflict after the land development has occurred and addressing the conflict prior to land development. The strategies will vary depending upon the situation at hand.

The important first step is to identify the range of alternative concepts for addressing these issues. It is also important to consider doing nothing—maintaining the status quo, absorbing the losses, and rebuilding in much the same way as before incurring the damage. While this may not seem like a solution, maintenance of the status quo often happens because it is easy to simply continue to do things as they were done in the past. If this alternative is chosen, it should be consciously selected as opposed to being selected by default or due to inertia. In other words, we need to own our decisions rather than lapse into them.

**Structural Solutions**

In the context of the upper Texas Coast after Ike, there has been great interest in levees as a mechanism for separating land development from the damaging surge. At this time, there are roughly 1.2 million people living in coastal hurricane evacuation zones that may need to evacuate during a given storm event. Additionally, there are major industrial and employment centers in areas that are likely to be impacted by major surge tides, including the industrialized Houston Ship Channel, the Bayport industrial complex, the NASA complex in the Clear Lake area, and the University of Texas Medical Center in Galveston.

In an attempt to protect these areas structurally, several different levee concepts have been identified at this time. The first is a linear levee running for almost 100 miles down the coast that has been termed the "Ike Dike." Although various alternative alignments have been shown, the levee system shown in figure 12.1 starts near High Island and is proposed to extend

down the Bolivar Peninsula and across the pass between Bolivar and Galveston Island to connect with the Galveston sea wall that was built after the hurricane of 1900. In order to cross the pass, which is known as "Bolivar Roads," a large gate structure will be built that will remain open much of the time and will be closed only when a storm approaches. On the south side of Bolivar Roads, the Ike Dike connects with the existing sea wall and then continues southward, crossing San Luis Pass and terminating somewhere on Follets Island, the peninsula extending northward from Freeport. Again, a gate structure will be utilized at San Luis Pass. Any cost estimates are premature, but they are certain to exceed several billion dollars and could be tens of billions of dollars. Variations on this alignment that have been publicly discussed include extending the levee all the way south to the existing Freeport levee system as well as extending northward into Jefferson and even potentially Orange counties. Regardless of the proposed configuration, the Ike Dike is a very large structural project.

There are many economic, environmental, and social impact issues associated with Ike Dike. Concerns range from the dollar cost to disruption of tidal circulation between Galveston Bay and the Gulf of Mexico to accelerated erosion and loss of beaches. On the other hand, this one alternative

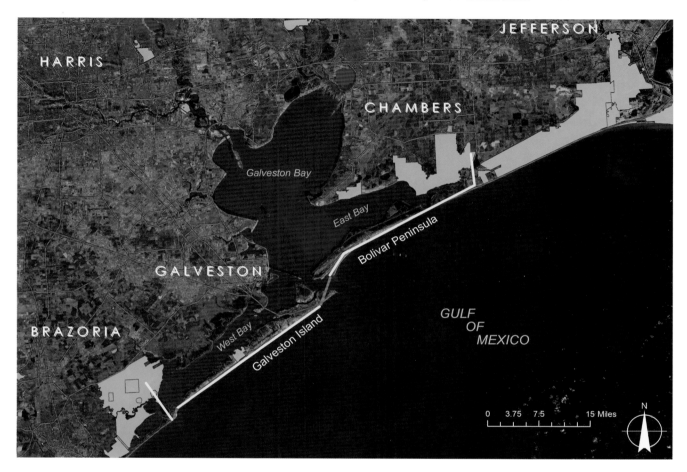

HARRIS

JEFFERSON

CHAMBERS

Galveston Bay

East Bay

Bolivar Peninsula

GALVESTON

GULF
OF
MEXICO

West Bay

Galveston Island

BRAZORIA

0    3.75    7.5         15 Miles

N

offers the promise of protection of much of the Houston-Galveston region, arguably providing as much as a $100 billion in avoided flood damages while costing tens of billions of dollars. This alternative is highly controversial and would alter the landscape of the upper Texas Coast for at least the next century. This is a very large alternative with associated big issues. Given current budgetary issues, cost alone may disqualify this alternative.

One alternative to the Ike Dike is a relatively simple levee and gate structure across the Houston Ship Channel at the Hartman SH 146 Bridge. As shown in chapter 5 (see fig. 5.10), such a gate structure would be erected across the point where the Houston Ship Channel/San Jacinto River complex enters Galveston Bay. The 25-foot contour line exists near the ship channel on both the north and south bank at this location, providing the ability to

Figure 12.1. The Ike Dike, extending the length of Galveston Island and the Bolivar Peninsula, has been proposed to protect the Houston/Galveston area. Courtesy Rice University Archives.

Figure 12.2. Upper west side of Galveston Bay protected by a levee.

create a 25-foot barrier with a set of gates similar to those proposed for the Ike Dike. This alternative has the ability to provide upwards of 50 percent of the economic benefits of the Ike Dike at a fraction of the cost and with virtually none of the environmental impacts, certainly making it worth studying in much greater detail. There are major issues that must be understood with this alternative, including the hydrologic issues of stopping the surge and then opening the gates to allow flood waters to drain, as well as the costs. However, this is a structural alternative that might have significant advantages on several levels.

In figure 12.2, one of a number of alternatives is shown for structurally protecting the upper west side of Galveston Bay. In this alternative, a levee is shown extending along the right-of-way of SH 146 from the Houston Ship Channel crossing southward to the existing Texas City levee system. The benefit of this alternative is that it could utilize the existing right-of-way of SH 146, keeping the cost of land acquisition down. With a location at this point, the levee would exclude portions of certain existing municipalities, including Morgan's Point, LaPorte, Shoreacres, Seabrook, Pasadena, Kemah, and San Leon, certainly creating certain social impacts. These social impacts could be mitigated by buy-out programs for flooded homes or, over time, houses in this unprotected area could be elevated and flood-proofed. It would also be possible to run a levee along the shoreline, but a similar concept was strenuously opposed when proposed by the USACE in the 1970s. As a general proposition, there is no easy structural fix for the problem of extensive existing development in this highly vulnerable part of the bay.

In figure12.3, one of several alternative levee alignments is shown for Galveston Island. At this time, the existing seawall protects the Gulf shoreline to an elevation of approximately 15 feet and proved generally adequate for Ike (which was not a worst case storm). However, Galveston experienced significant flooding from the bay side which is unprotected. Of particular importance is the protection of the University of Texas Medical Branch (UTMB) which was extensively flooded by Ike. In figure 12.3, a levee is shown by the dotted line extending adjacent to UTMB and the Strand and continuing westward and then crossing Offats Bayou and connecting back to the sea wall. Under this alignment, only one gate structure would be required. If protection of some or all of Pelican Island were desired, then additional gates would need to be installed, significantly increasing the cost of the levee.

**Nonstructural Alternatives**

Nonstructural alternatives are quite different from structural alternatives and are not particularly well understood or well developed compared to other flood control solutions. The starting point of a nonstructural flood control alternative is to identify the area at risk and understand how it is and/or could be developed in order to minimize the potential conflict between development and flooding. Extensive inundation occurred miles inland from Galveston Bay east to Grand Isle, Louisiana. Although there was major damage immediately adjacent to Galveston Bay, the majority of the development in the extremely vulnerable west side of Galveston Bay was not hit very hard by Ike. Instead, the bulk of the damage occurred on Bolivar, in the City of Galveston, and in certain areas along the west shore that were exposed to the "fetch" coming from the north down the bay system from the backside of Ike. For the most part, the upper west side of the bay was spared.

Nonstructural alternatives are what the name implies—they are solutions that do not primarily rely on constructed solutions to attempt to contain the natural feature. Instead, approaches are designed based upon an understanding of the natural features such as storm surge and attempt to work with those features

Galveston Island Levee and Alternatives

0   967  1,933    3,867      5,800      7,700
Feet

Figure 12.3. Galveston Island protected by a levee.

rather than attempting to "control" the storm waters. As such, these alternatives would respect and work with inundation, rather than attempting to prevent inundation. These alternatives also have the benefit of generally having less ecological impact than many structural alternatives.

Tremendous amounts of water flooded the land areas east of Galveston Bay, extending to Grand Isle, Louisiana, and beyond, as mentioned in chapter 2 (see fig. 2.5). For the most part, there was relatively little damage done in the majority of these low-lying areas due to the absence of development. In fact, storm surge drained off these land areas for several days af-

ter Ike made landfall. These low-lying lands provide a service by storing flood waters with little economic damage. The natural environmental system, while negatively impacted, has evolved to absorb these occasional storm surges and can survive and eventually thrive after inundation. So, an initial non-structural strategy would be to protect and conserve these natural areas that hold extensive volumes of flood water.

One method of conserving and protecting much of these lands would be to establish some form of land trust or protected-lands strategy. Another concept could be to establish some variation of a national park along the upper Texas Coast, potentially modeled after some of the innovative concepts developed for national recreation areas in other areas of the United States. Over 4 million persons live within 60–90 miles of the East and West Bay shorelines, which are some of the best recreational areas in the United States, and offer world-class bird-watching, fishing, hunting, crabbing, and kayaking opportunities, as well as also holding extensive volumes of storm surge. Such a system could be developed in conjunction with local governmental and private sector interests to develop a partnership for development of a thriving and yet sustainable coastal economy.

A potential location for a national recreation area is shown in figure 12.4, which shows the low-lying areas of the bay in shades of green and the major access and ingress from existing developed areas. Here, there is the potential that a resilient economy could be developed for these low-lying areas, one that works in harmony with the natural system rather than against it. The bird-watching and kayaking potential of these coastal assets is incredible, offering access to the estimated 90 million persons who spend their leisure in this manner. As such, the nonstructural alternative here is protecting these lands as much through resilient economic development as in setting the lands aside. Additionally, there are benefits that can be obtained

from coordinating carbon sequestration efforts with preservation and reconstruction of saltwater wetlands, native prairies, and forested lands. Over a number of years, these sequestration values will rise, offering additional support for easement and/or fee-simple purchase.

The nonstructural "natural strategy" explained above could be implemented as a partnership between the federal, state, and local governments along with non-governmental organizations and private landowners. The proposed national recreation area (NRA) would be created by an agreement that would set out the rights and responsibilities of the parties comprising the NRA. There would be no additional regulation on nonparticipants and no unwilling participation. The focus of the National Recreation Area would be on the low-lying areas where virtually no land development interest exists at this time. For the most part, this national recreation area strategy would seek to work with the lands that are most desirable from an ecological perspective and less desirable from a development standpoint. Over time, funds could be set aside to buy property from willing buyers. As shown in figure 12.4, the existing communities of Freeport/Surfside, Galveston, and High Island would be the gateways to these new as well as traditional recreational uses.

168

Figure 12.4. National Recreation Area study area, showing major access routes and gateway communities. © 2011 Bryan Carlile, Beck Geodetix

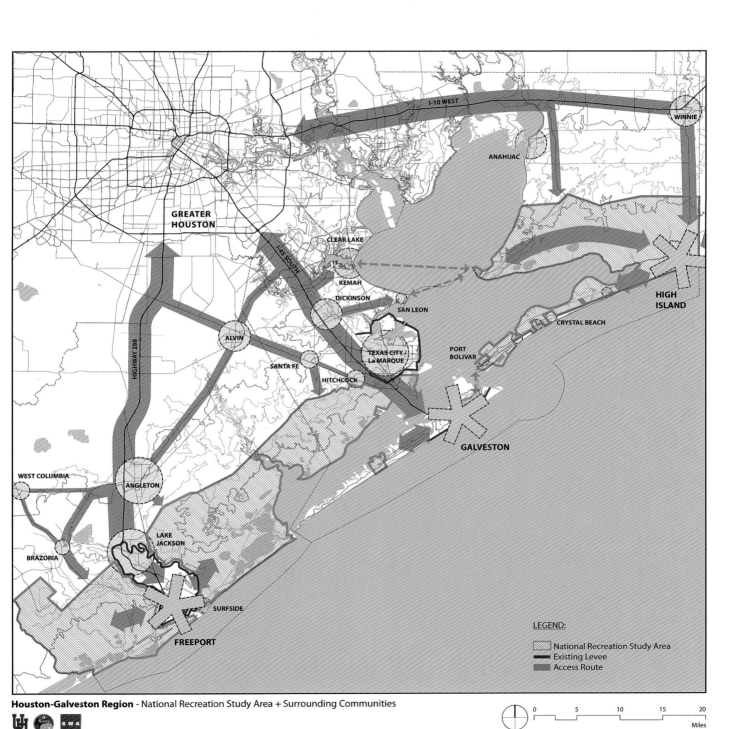

**Houston-Galveston Region** - National Recreation Study Area + Surrounding Communities

LEGEND:

National Recreation Study Area
Existing Levee
Access Route

0    5    10    15    20
Miles

From a nonstructural perspective, the developed upper west side of Galveston Bay presents its own set of daunting problems. This area is developed yet is a very dangerous place to be when a hurricane makes landfall. Areas along the coast have been classified as hurricane evacuation zones (HEZs) as discussed in chapter 8 (see fig. 8.4). There is great concern about whether or not the 1.2 million people living in this zone can actually be evacuated in a timely manner. Additionally, development projections from the Houston Galveston Area Council project that approximately 700,000 people are anticipated to move into the HEZ over the next 25 years (HGAC 2007). There is little optimism that all of these people can be evacuated from these highly dangerous areas in a timely manner.

The development of these evacuation zones is a major concern. There is not much that can be done in the short term about the existing residents of the HEZ but there are certain actions that can be taken with regard to those who have not yet moved into the area. For one thing, no information is currently provided to those who are buying property in the HEZ about the location of their property. No one is told at closing that they are moving into a HEZ and it is doubtful that most people moving in from other parts of the United States and many people from the Houston region are aware of the

substantial risk posed by hurricanes. One nonstructural alternative would be to provide accurate information about the fact that property is located in the HEZ. Perhaps more importantly, information should be provided about the surge tide risk associated with various storm events. A potential disclosure mechanism could be a diagram provided to a potential homebuyer to alert them about the extent of surge flooding that would be expected with a particular storm and/or surge event (figs. 12.5a, b).

An alternative type of disclosure could be placement of markers within the community to identify the depth of inundation that is possible from the storm surge associated with variously sized storms. The concept underlying the potential effectiveness of either of these approaches is disclosure of information that is not generally known to many members of the public. If you are new to the Gulf Coast, you simply may not know about the HEZ or surge tides.

Next among the nonstructural alternatives would be altering the Flood Insurance Rate Map (FIRM) that is currently in effect in the upper west side of Galveston Bay. FIRMs display the 100-year flood plain that is used for regulatory purposes. However, with the exception of areas immediately adjacent to the bay, hurricane surge tide flooding is largely not considered in these maps which instead concentrate on stream

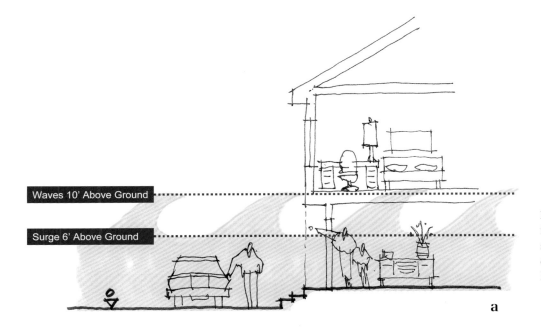

Waves 10' Above Ground

Surge 6' Above Ground

a

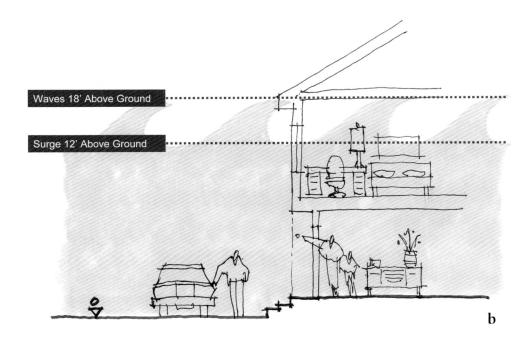

Waves 18' Above Ground

Surge 12' Above Ground

b

Figure 12.5a–b. 10-foot (a) and 18-foot (b) surge and wave heights depicted with respect to homes in coastal areas.

and bayou flooding. A comparison of the area covered by the 100-year flood plain versus the area inundated by a 20.4-foot surge tide is shown on figure 12.6. For the most part, the building slab elevation required in the flood plains is only about 11 or 12 feet rather than the 20.4 feet associated with a reasonable worst case surge tide. At this time, an effort is underway to change the FIRMs for the Texas coast. This effort is anticipated to lead to the publication of new maps in 2012.

Yet another nonstructural alternative involves the development and maintenance of an extensive flood and surge warning system that could be accessed 24/7 by members of the public as well as governmental officials. Such a system is an essential part of evacuation planning, but is also a key aspect of returning to an area that may have been struck in part or in whole by incoming storm. A conceptual flood warning system is shown in figure 12.7. Here, data is collected from buoys in the Gulf and along the bayfront as well as from rain gauges, radar, and stream gauges. As the storm approaches, real-time information can be displayed showing surge heights, rainfall levels, and areas of expected flooding. After passage of the storm, information about residual stream flooding, street, and power conditions could be relayed. If nothing else, such a system would aid in the orderly evac-

uation and return of residents to the evacuation zone. This system could be operated in conjunction with certain flood-proofing alternatives as is the case with the Texas Medical Center. The technology for this system is basically in existence. All that remains is to gain access to the various sources of data, supplement the data where necessary, and make this information available via the internet.

Another non-structural opportunity would be along the Houston Ship Channel where a significant amount of infrastructure, including hazardous chemical storage and treatment facilities, exists. For the most part, these facilities have been designed to the Federal Emergency Management Agency (FEMA) FIRM 100-year flood plain standard. A quick review of the FEMA maps show that the 100-year flood plain along the Houston Ship Channel is in the 12- to 14-foot range. Under the rules of the US Environmental Protection Agency and the Texas Commission on Environmental Quality (TCEQ), hazardous waste storage facilities and wastewater treatment facilities must be protected to the 100-year flood elevation. However, many facilities, such as storage tanks, are not protected from the 100-year flood under regulation. Perhaps more importantly, up the Houston Ship Channel there is a substantial difference between the mapped 100-year flood plain and the

Figure 12.6. Regions in the Houston/Galveston area inundated by 20.4-foot storm surge. © 2011 Bryan Carlile, Beck Geodetix.

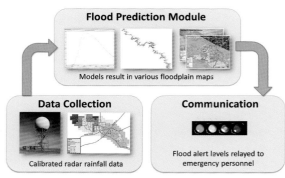

Figure 12.7. A conceptual flood alert system, like the Rice/TMC FAS, that could be employed in coastal communities. Courtesy Rice University Archives.

flooding that would have been associated with Ike had it made landfall farther west or been larger. Had Ike been 30 percent larger and made landfall at Point 7 (see fig. 5.6) many facilities handling hazardous waste in addition to wastewater treatment plants would have been inundated as discussed in chapter 10. If the channel is not structurally protected, then individual facilities will need to take some type of action to ensure that they are protected from the surge. This can be accomplished by relocating these facilities, by levee construction, or perhaps by flood-proofing. In order to implement such a requirement as a nonstructural control, the TCEQ could amend its regulations to require protection from a surge event in a specific area such as the Houston Ship Channel.

And finally, a separate nonstructural strategy involves flood insurance in the most vulnerable areas. To the extent that substantial development continues to occur along the barrier island, the major issue is who should bear the risk of this development activity. This question, in turn, raises issues about the role of the Federal Flood Insurance Program and coastal development. FEMA provides flood insurance in areas where the local government agrees to regulate building elevation. However, there are some areas where flood insurance is unavailable. Portions of Galveston Island and Bolivar remain

eligible for flood insurance while others do not. The potential to further restrict where flood insurance is available on barrier islands and peninsulas still exists. Rather than preventing development, private landowners would be able to build what they wished. However, by restricting flood insurance, the financial risk would be assumed by the homeowner rather than taxpayers. In a state that protects landowner rights and eschews government intervention, such an approach should be consistent with stated philosophies.

## No Action Alternative

In addition to all the alternatives discussed, there is the alternative of doing nothing. To some extent, a full no action alternative is unlikely to occur due to the fact that building codes either have been or are being changed in many coastal areas and many coastal builders are choosing to voluntarily increase flood slab elevations and construction specifications. However, in a larger sense, no action implies that hundreds of thousands more people will move into the HEZ without forewarning and without adequate understanding of the risk. No action means that no structural protection will be erected although individual property owners might choose to either build a levee to protect their property or otherwise attempt to flood-proof their property. Contracts have been let to redraw the FEMA FIRMs for the coastal area, so these maps will be changed over the next several years. However, there may be no coherent policies or approaches. Similarly, if the decision is made to pursue one or more of the structural alternatives in the longer term, no action is likely to result in the short term and potentially in the long term if that structural alternative cannot be funded or authorized. Overall, no action seems an undesirable yet somewhat likely outcome of the current situation. By drawing attention to this no action alternative, we are intending to remind everyone in the region that we know what can happen—and likely will happen—not if but when a storm as bad or worse than Ike makes a direct hit on our area. While doing nothing sounds ridiculous to some, it is more often than not a reality that we all need to recognize and own. If we fail to take significant action, we will have selected the worst alternative.

## Conclusion

What has been laid out in the bulk of this chapter are a number of various structural and nonstructural alternatives including no action. The major task before the community is to evaluate these various approaches

Table 12.1. Example sustainability matrix for Houston/Galveston area.

| Project | Environmental Impacts to Mainland and Bay | [Full] Economic Costs and Benefits | Social Impacts |
|---|---|---|---|
| No Action | | | |
| Ike Dike | | | |
| Ship Channel Dike | | | |
| Galveston Levee | | | |
| National Recreation Area | | | |
| Stronger Land-Use Controls | | | |

(and others that may be developed) from an economic, social, and ecological perspective. We need to know how well these alternatives will work, how much they will cost, and the benefits and harms that will be associated with the construction, operation, and maintenance of each. Essentially, a detailed matrix will need to be filled out similar to table 12.1. When the matrix is complete, the Houston-Galveston region will be well on its way to implementing a plan to address regional vulnerability to hurricane storm surge.

# Glossary

**A**

Abutment: the structural element, typically constructed of concrete or timber, at either end of a bridge that supports the ends of the bridge.

Advanced Circulation (ADCIRC) Model: the model used to map coastal circulation and storm surge in the event of an approaching hurricane.

Approach (to a bridge): the part of the bridge that carries traffic from the land onto the bridge.

**B**

Bascule Section: a movable span that rotates to raise one end vertically. A bridge may have one or two sections that meet in the center to allow large ships to pass beneath the bridge.

Bathymetry: a measurement of the depth of water in coastal areas. Shallow bathymetry contributes to heightened storm surge.

Bent Beam: a horizontal structural element (beam) that supports the bridge span, and transfers loads from the super structure to the columns. It is also called a bent cap.

Box-Girder: a bridge girder element that is box-like in shape. It may be hollow or solid.

**C**

Contraflow: the reversal of inbound freeway lanes to allow efficient outbound traffic flow during an emergency event.

Council on Environmental Quality (CEQ): established within the Executive Office of the President under the National Environmental Policy Act (NEPA) to coordinate federal environmental efforts and work with agencies and other White House offices to develop environmental policies and initiatives.

**D**

Deck: the roadway portion of a bridge.

Deck Bridge: a bridge in which the supporting members are all beneath the roadway.

Deck Uplift: is caused by storm surge when the force attributed to wave loading exceeds the weight of the deck, lifting it off of its supports.

Digital Hybrid Reflectivity (DHR): monitors average accumulated precipitation in a basin at a resolution of 1 degree by 1 km.

Directed Evacuation: *see Mandatory Evacuation*

Disaster Recovery Centers (DRC): a facility or mobile office where one can go for information about FEMA or other disaster assistance programs.

## E

El Niño- Southern Oscillation (ENSO): a warm phase associated with high sea surface temperatures off the coast of Peru, low atmospheric pressure over the eastern Pacific, and increased vertical wind shear over the Atlantic.

Emergency Management Coordinator (EMC): leads a staff of trained professionals to determine a method for communication within a community and course of action to be taken prior to, during, and after an emergency event.

Enhanced Fujita Tornado Damage Scale: a scale from EF0-EF6 indicating the most intense damage caused by a tornado event.

## F

Federal Emergency Management Agency (FEMA): created in 1979 to perform all federal activities related to disaster mitigation and emergency preparedness, response and recovery.

Flash Flood Monitoring and Prediction (FFMP): a collaborative effort involving the National Weather Service (NWS), the National Severe Storms Laboratory (NSSL), and the National Center for Atmospheric Research (NCAR) to improve the accuracy and timeliness of warnings issued by NWS forecasters, through the development of automated warning guidance. (National Weather Service 2009)

Flash Flood Warning: issued to inform the public, emergency management agencies, and other cooperating agencies that flash flooding is in progress, imminent, or highly likely. (National Weather Service 2009)

Flash Flood Watch: issued to indicate current or developing hydrologic conditions that are favorable for flash flooding in and close to the watch area, but the occurrence is neither certain or imminent. (National Weather Service 2009)

Flash Flood: caused by heavy or excessive rainfall in a short period of time, generally less than 6 hours. (National Weather Service 2009)

Flood Insurance Rate Map (FIRM): an official map released by FEMA that indicates probabilistic areas of flooding to be used during urban planning and development.

Flood: an overflow of water onto normally dry land; the inundation of a normally dry area caused by rising water in an existing waterway; ponding of water at or near the point where the rain fell. (National Weather Service 2009)

Floodplain: the portion of a river valley that has been inundated by a river during historic flooding. (National Weather Service 2009)

## G

Geographic Information System (GIS): integrates hardware, software, and data for capturing, managing, analyzing, and displaying all forms of geographically referenced information.

Girder: a horizontal structure member supporting vertical loads by resisting bending.

## H

Harris County Flood Control District (HCFCD): a government agency developed to address the affects of flooding in the Harris County, Texas by creating flood damage reduction plans and maintaining flood control infrastructure.

Houston Galveston Area Council (HGAC): the region-wide, voluntary association of local governments in the Gulf Coast Planning region of Texas. Its service area is 12,500 square miles and contains more than 6 million people.

Hurricane Evacuation Zones (HEZs): designated coastal areas that may be advised to evacuate during the approach of a hurricane.

Hurricane: a tropical cyclone in the Atlantic, Caribbean Sea, Gulf of Mexico, or eastern Pacific that has maximum 1-minute sustained surface winds greater than 73 mph.

## I

Ike Dike: a proposed extension of the existing levee system on Galveston Island. The actual length and location of the Ike Dike has not yet been determined.

Incident Commander (IC): approves communications with the public during the approach and onset of an emergency.

## L

La Niña: a cold phase associated with low sea surface temperatures in the eastern Pacific, low atmospheric pressure over the western Pacific, and decreased vertical wind shear over the Atlantic.

Light Detection and Ranging (LIDAR): topographic information for a given area determined by measuring the distance to the ground relative to a given datum using aircraft mounted with optical remote sensing technologies (lasers).

## M

Mandatory or Directed Evacuation: an announcement that danger to a community is imminent and those community members must leave the area.

Mesoscale Convective Systems (MCS): a complex of thunderstorms that become organized and persist for several hours or more. They may be round or linear in shape, and include systems such as tropical cyclones or squall lines, among others. (National Weather Service 2009)

Metropolitan Statistical Area (MSA): an urban area with a population one million or more and separate component areas meeting specific statistical criteria and supported by local opinion.

Millibar (Mb, Mbar): a unit of atmospheric pressure equal to 1/1000 bar.

Mitigation: the effort to reduce loss of life and property by lessening the impact of disasters.

## N

National Environmental Policy Act (NEPA): a US environmental law established in 1970 that created a US national policy to promote the enhancement of the environment and that established the President's Council on Environmental Quality (CEQ).

National Hurricane Center (NHC): an organization that continuously watches tropical cyclones activity in the Atlantic, Caribbean, Gulf of Mexico, and eastern Pacific. It distributes hurricane watches and warnings to the general public and prepares and distributes marine and military advisories.

National Oceanic and Atmospheric Association (NOAA): a scientific agency within the US Department of Commerce focused on the conditions of the oceans and the atmosphere.

National Recreation Area (NRA): a designated or protected recreational area managed by the federal government.

The National Weather Service (NWS): provides weather, hydrologic, and climate forecasts and warnings and is the official voice of the United States for issuing warnings during life-threatening weather situations.

Next Generation Radar (NEXRAD): a National Weather Service network of about 140 Doppler radars operating nationwide. (National Weather Service 2009)

## P

Public Information Officer (PIO): in charge of developing a format and determining the functionality of communication with the public.

## R

Rip-rap: a material, typically rocks or boulders, placed on the slope in front of a bridge abutment to protect it from the effects of scour.

River Forecast Centers (RFC): prepare river and flood forecasts and warnings and provide hydrologic guidance to Weather Service Forecast offices and Weather Forecast offices.

## S

Saffir-Simpson Hurricane Scale: a scale of 1 to 5 based on the maximum sustained surface wind speed associated with a tropical cyclone.

Scour: the removal of sediments from around bridge abutments or piers by water flow compromising the integrity of a structure.

The Sea, Lake, and Overland Surges from Hurricanes (SLOSH) Model: a model run by the National Hurricane Center (NHC) to estimate storm surge heights during a tropical storm.

Sea Surface Temperatures (SSTs): the mean temperature of the ocean in the upper few meters.

Severe Storm Prediction, Education and Evacuation from Disasters (SSPEED) Center: established in 2007 as a university-based research and education organization. Led by Rice University, the Center organizes leading universities, researchers, emergency managers, and private and public entities to better address severe storm prediction and its impacts.

Shear Key or Shear Link: provides transverse support to a bridge superstructure during lateral loading helping to prevent deck unseating.

Shelter-in-Place: an advisory to citizens to remain where they are rather than to evacuate because there is no imminent danger to their location.

Simulcast: a "simultaneous broadcast" in which a message is broadcasted over more than one medium, or more than one service on the same medium, at the same time.

Small Business Administration (SBA): a government agency that provides support to small businesses.

Social Vulnerability (SV): the ability of a person or community to anticipate, cope with, resist, and recover from a disaster event.

Spalling: occurs after salt water enters concrete causing the surface to peel or flake off.

Spans: the horizontal space between two structural supports of a structure.

Storm Surge: is measured as the rise in sea level above the normal tide level in advance of and during a tropical cyclone.

Stratosphere: the layer of the earth's atmosphere above the troposphere, extending to about 50 km above the surface of the earth.

Substructure: all structural parts of a bridge that support the superstructure and are below the bridge deck base line. The main components are abutments or endbents, piers or interior bents, footings, and pilings.

Superstructures: all structural parts of the bridge on top of the substructure. The main components are the bridge deck, parapets, and railings.

SWAN Model: a computer model to compute irregular waves in coastal environments.

**T**

Texas Coastal Community Planning Atlas: an internet-based, spatial decision support system that allows users to identify and visualize critical issues related to numerous dimensions of development including environmental degradation, natural hazard risks, and significant changes in land use patterns.

Texas Department of Rural Affairs (TDRA): a state agency that ensures a continuing focus on rural issues, monitors governmental actions affecting rural Texas, researches problems and recommends solutions, and coordinates rural programs among state agencies.

Texas Department of Transportation (TxDOT): a state agency that works to provide safe, effective, and efficient movement of people and goods in the state of Texas through the construction and maintenance of the transportation systems of the state.

Timber Bridge: a wooden bridge generally used to span a creek or other relatively small crossings.

TranStar: a partnership of four state and local government agencies (TxDOT, Harris County, Metropolitan Transit Authority of Harris County, and the City of Houston) that is responsible for providing transportation and emergency management services to the Greater Houston region.

Tropical Depression: a tropical cyclone in which the maximum 1-minute sustained surface wind is 38 mph or less.

Tropical Storm Allison Recovery Project (TSARP): a joint study effort by FEMA and HCFCD to develop technical products that assist the local community in recovery from flooding during Tropical Storm Allison, and provide the community with a greater understanding of flooding and flood risks.

Tropical Storm: A tropical cyclone in which the maximum 1-minute sustained surface wind ranges from 39 to 73 mph.

**V**

Voluntary Evacuation: occurs when members of a community choose to leave an area because they fear the danger of sheltering-in-place. Voluntary evacuation may not be advised by public officials because it can increase traffic congestion and hinder mandatory evacuations.

**W**

Watershed: a contiguous land area that drains to a single outlet.

Weather Forecast Offices (WFO): are responsible for issuing advisories, warnings, statements, and short-term forecasts for its county warning area.

Wing Wall: an extension of the abutment up- and downstream that helps protect the bridge foundation from erosion.

World Meteorological Organization: an intergovernmental organization with a membership of 189 member states and territories specializing in meteorology, operational hydrology, and related geophysical sciences.

WSR-88D Doppler Radar: a weather surveillance radar.

**Z**

Zip-zone Map: a map showing evacuation zones identified by zip code.

# Sources

Arnold, L. A. and Gibbons, C. J. 1996. Impervious surface coverage-the emergence of a key environmental indicator *Journal of the American Planning Association* 62: 243–58.

Bedient, P., H. Wayne, and B. Vieux. 2008. *Hydrology and floodplain analysis.* 4th ed. Upper Saddle River, N.J: Prentice Hall.

Berg, R. 2009. *Tropical Cyclone Report: Hurricane Ike 1–14 September 2008.* Miami, Fla.: National Hurricane Center. http://www.nhc.noaa.gov/pdf/TCR-AL092008_Ike_3May10.pdf.

Beven, J. 2004. *Tropical Cyclone Report: Hurricane Frances 25 August—8 September 2004.* Miami, Fla.: National Hurricane Center. http://www.nhc.noaa.gov/pdf/TCR-AL062004_Frances.pdf.

Birkland, T. A., R. J. Burby, D. Conrad, H. Cortner, and W. K. Michener. 2003. River ecology and flood hazard mitigation. *Natural Hazards Review* 4(1):46–54.

Blaikie, P., T. Cannon, I. Davis, and B. Wisner. 1994. *At Risk: Natural hazards, people's vulnerability and disasters.* London: Routledge.

Blake, E., E. Rappaport, and C. Landsea. 2007. *The deadliest, costliest, and most intense United States tropical cyclones from 1851–2006.* NOAA Technical Memorandum NWS TPC-5. Miami, Fla.: National Hurricane Center. http://www.nhc.noaa.gov/pdf/NWS-TPC-5.pdf.

Bolin, R. 1986. Disaster impact and recovery: A comparison of black and white victims, *International Journal of Mass Emergencies and Disasters* 4: 35–50.

Bolin, R. and P. Bolton. 1986. *Race, religion, and ethnicity in disaster recovery.* Program on Environment and Behavior, Monograph 42. Boulder, Colo.: Institute of Behavioral Science, University of Colorado.

Bolin, R. and L. Stanford. 1991. Shelter, housing and recovery: A comparison of US disaster. Disasters 15: 24–34.

Brody, S.D., V. Carrasco, and W. E. Highfield. 2006. Measuring the adoption of local sprawl reduction planning policies in Florida. *Journal of Planning Education and Research* 25: 294–310.

Brody, S. D., S. E. Davis, III, W. E. Highfield, and S. Bernhardt. 2008. A spatial-temporal analysis of wetland alteration in Texas and Florida: thirteen years of impact along the coast. *Wetlands* 28: 107–16.

Brody, S. D., W. E. Highfield, H. Ryu, and L. Spanel-Weber. 2007c. Examining the relationship between wetland alteration and watershed flooding in Texas and Florida. *Natural Hazards* 40: 413–28.

Brody, S.D., J. E. Kang, S. Zahran, and S.P. Bernhardt. 2009. Evaluating local flood mitigation strategies in Texas and Florida. *Built Environment* 35: 492–515.

Brody, S.D., S. Zahran, W. E. Highfield, H.

Grover, and A. Vedlitz. 2007b. Identifying the impact of the built environment on flood damage in Texas. *Disasters* 32(1): 1–18.

Brody, S.D., S. Zahran, P. Maghelal, H. Grover, and W. Highfield. 2007a. The rising costs of floods: examining the impact of planning and development decisions on property damage in Florida. *Journal of the American Planning Association* 73: 330–45.

Bullock, A. and M. Acreman. 2003. The role of wetlands in the hydrological cycle. *Hydrology and Earth System Sciences* 7: 358–89.

Bunya, S., J. C.Dietrich, J. J.Westerink, B. A. Ebersole, J. M. Smith, J. H. Atkinson, R. Jensen, D. T. Resio, R. A. Luettich, C. Dawson, et al. 2010. A high resolution coupled riverine flow, tide, wind, wind wave and storm surge model for southern Louisiana and Mississippi: part I-model development and validation. *Monthly Weather Review* 138(2): 345–77.

Burby, R. J. 2002. *Flood insurance and floodplain management: The US experience.* Chapel Hill, N.C.: Department of City and Regional Planning, University of North Carolina.

Carter, R.W. 1961. Magnitude and frequency of floods in suburban areas. *USGS Professional Paper* 424-B: 9–11.

CenterPoint Energy. 2010 Hurricane Ike: Like many things, bigger in Texas. http://www.CenterPointenergy.com/newsroom/stormcenter/ike/.

Crossett, K., T. J. Culliton, P. Wiley, and T. R. Goodspeed. 2004. *Population Trends Along the Coastal United States, 1980–2008.* Coastal Trends Report Series. Washington, D.C.: National Oceanic and Atmospheric Administration, National Ocean Service. http://oceanservice.noaa.gov/programs/mb/pdfs/coastal_pop_trends_complete.pdf

Cutter, S. L. 1996. vulnerability to environmental hazards. *Progress in Human Geography* 20: 529–39.

Dahl, T. E. 2000. *Status and trends of wetlands in the conterminous United States 1986–1997.* Washington, D.C.: US Fish and Wildlife Service.

Dash, N., Peacock, W.G. and Morrow, B. 1997. And the Poor Get Poorer: A Neglected Black Community. In *Hurricane Andrew: Ethnicity, gender and the sociology of disaster,* ed. W. G. Peacock, B. H. Morrow, and H. Gladwin, 206–25. London: Routledge.

Dawson, Clint and Jennifer Proft. Hurricane Ike ADCIRC and SWAN Data. University of Texs at Austin. Files sent to author 23 June 2010.

DesRoches, R., ed. 2006. *Hurricane Katrina: Performance of transportation systems.* Technical Council on Lifeline Earthquake Engineering Monograph No. 29. Baltimore: American Society of Civil Engineers.

Deyle, R. E., S. P. French, R. B. Olshansky, and R. G. Paterson. 1998. Hazard assessment: The factual basis of planning and mitigation. In *Cooperating with nature: Confronting natural hazards with land use planning,* ed. R. J. Burby, 119–66. Washington, DC: Joseph Henry Press.

Dietrich, J. C., S. Bunya, J. J. Westerink, B.

A. Ebersole, J. M. Smith, J. H. Atkinson, R. Jensen, D. T. Resio, R. A .Luettich, C. Dawson, et al. 2010. A high-resolution coupled riverine flow, tide, wind, wind wave and storm surge model for southern Louisiana and Mississippi: part II-synoptic description and analysis of hurricanes Katrina and Rita. *Monthly Weather Review* 138(2): 378–404.

Doswell, C. 1978. Severe storms. In *Encyclopedia of atmospheric sciences*, MS-366. Norman, Okla: National Oceanic and Atmospheric Administration, National Severe Storms Laboratory. http://www.flame.org/~cdoswell/publications/Severe_Storms_EncyHolton.pdf.

Dunne, T. and L. B. Leopold. 1978. *Water in environmental planning.* New York: Freeman.

EIA (Energy Information Administration). 2008. *Impact of the 2008 hurricanes on the natural gas industry.* US Department of Energy, Office of Oil and Gas. http://www.eia.doe.gov/ . . . /natural . . . /nghurricanes08/nghurricanes08.pdf.

Enarson, E. 1999. Violence against women in disasters: A study of domestic violence programs in the US and Canada. *Violence Against Women* 5: 742–68.

Enarson, E. and B. H. Morrow, eds. 1998. *The gendered terrain of disaster: Through women's eyes.* Westport, Conn.: Praeger.

Enarson, E. and Morrow, B.H. 1997. A gendered perspective: The voices of women. In *Hurricane Andrew: Ethnicity, gender and the sociology of disasters,* ed. W. G. Peacock, , B. H. Morrow, and H. Gladwin, 116–40. London: Routledge.

FEMA (Federal Emergency Management Agency). 1996. *Guide for all-hazard emergency operations planning.* SLG 101. Washington, D.C. http://www.fema.gov/pdf/plan/slg101.pdf

FEMA (Federal Emergency Management Agency). 2008. *Hurricane Ike impact report.* Washington, D.C.: Department of Homeland Security. http://www.fema.gov/pdf/hazard/hurricane/2008/ike/impact_report.pdf

FEMA. 2010. Tropical Storm Allison recovery project (TSARP). FEMA and Harris County Flood Control District. http://www.tsarp.org/.

Fleming, J., C. Fulcher, R. Luettich, B. Estrade, G. Allen, and H. Winer. 2008. A real time storm surge forecasting system using ADCIRC. In *Estuarine and Coastal Modeling (2007)*, ed. M. L. Spaulding, 893–912. Baltimore: American Society of Civil Engineers.

Florida Disaster. 2010. *Statewide emergency shelter plan.* Tallahassee, Fla.: Florida Division of Emergency Management. http://www.floridadisaster.org/Response/engineers/SESPlans/2010SESPlan/.

Fothergill, A., E. Maestas, and J. D. Darlington. 1999. Race ethnicity and disasters in the US: A review of the literature. *Disasters* 23(2):156–73.

Fothergill, A. and L. A. Peek. 2004. Poverty and disasters in the United States: A review of recent sociological findings. *Natural Hazards* 32(1): 89–110.

Frech, M. 2005. Flood risk outreach and the public's need to know. *Journal of Contemporary Water Research and Education* 130: 61–69.

Harris, D. Lee. 1963. *Characteristics of the*

*hurricane storm surge.* US Department of Commerce Weather Bureau Technical Paper No. 48. Washington D.C.: US Government Printing Office.

Highfield, W. E. and S. D. Brody. 2006. The price of permits: Measuring the economic impacts of wetland development on flood damages in Florida. *Natural Hazards Review* 7(3): 23–30.

HGAC (Houston-Galveston Area Council). 2007. *Bridging our communities: 2035 Houston-Galveston Regional Transportation Plan.* http://www.h-gac.com/taq/plan/default.aspx

Kennedy, A. B., U. Gravois, B. C. Zachry, J. J. Westerink, M. E. Hope, J. C. Dietrich, M. D. Powell, A. T. Cox, R. A. Luettich Jr., and R. G. Dean. 2011. Origin of the Hurricane Ike forerunner surge. *Geophysical Research Letters* 38: L08608.

Knabb, R., J. Rhome, and D. Brown. 2006a. *Tropical Cyclone Report: Hurricane Katrina 25–30 August 2005.* Miami, Fla.: National Hurricane Center. http://www.nhc.noaa.gov/pdf/TCR-AL122005_Katrina.pdf.

Knabb, R., J. Rhome, and D. Brown. 2006b. *Tropical Cyclone Report: Hurricane Rita 18–16 September 2005.* Miami, Fla.: National Hurricane Center. http://www.nhc.noaa.gov/pdf/TCR-AL182005_Rita.pdf>.

Lawrence, M. and H. Cobb. 2005. *Tropical Cyclone Report: Hurricane Jeanne 13–28 August 2004.* Miami, Fla.: National Hurricane Center, Print. http://www.nhc.noaa.gov/pdf/TCR-AL112004_Jeanne.pdf

Lindell, M. K. and R. W. Perry. 1992. *Behavioral foundations of community emergency planning.* Washington, D.C.: Hemisphere.

Lindell, M.K. and R.W. Perry. 2004. *Communicating environmental risk in multiethnic communities.* Thousand Oaks, Calif: Sage Publications.

Little, J. E. 2006. *Task Force on Evacuation, Transportation, and Logistics: Final Report to the Governor.* Austin, Tex. ftp://ftp.txdps.state.tx.us/dem/hurr/GovHurrTaskForceReport.pdf.

Merrell, William J., Lydda Graham Reynolds, Andres Cardenas, Joshuya R. Gunn, and Amie J. Hufton. 2010. The Ike Dike: A Coastal Barrier Protecting the Houston/Galveston Region from Hurricane Storm Surge. In *Macro-Engineering Seawater in Unique Environments: Arid Lowlands and Water Bodies Rehabilitation.* Eds. Viorel Badescu and Richard B. Cathcart, 691–716. Heidelberg, Berlin: Springer Verlag.

Merrill, Brian D., PE. 2010. "SH 87 at Rollover Pass—Bridge Inspection Summary." Report sent to author, 15 January.

Mileti, D. S. 1999. *Disasters by design: A reassessment of natural hazards in the United States.* Washington D. C.: Joseph-Henry Press.

Mitch, W. J. and J. G. Gosselink. 2000. *Wetlands.* 3rd ed. New York: John Wiley & Sons.

Mitchell, J. K., ed. 1989. *The long road to recovery: Community response to industrial Disaster.* Tokyo: United Nations University Press.

Morrow, B. H. 1999. Identifying and mapping community vulnerability. *Disasters* 23(1): 1–18.

Mosqueda, G., K. A. Porter, J. O'Connor, and P. McAnany. 2007. Damage to engineered buildings and bridges in the wake of Hurricane Katrina. In *Forensic engineering,* ed. E. C. Stovner, 1–11. Baltimore: American Society of Civil Engineers.

Murray-Tuite, P. M., and H. S. Mahmassani. 2004. Transportation network evacuation planning with household activity interactions. *Transportation Research Record* 1894: 150–59.

National Weather Service. 2009. National Weather Service Glossary. http://www. weather.gov/glossary/.

NRC (National Research Council). 2000. *Risk Analysis and uncertainty in flood damage reduction studies.* Washington, DC: The National Academies Press.

NRC (National Research Council). 2006. *Facing hazards and disasters: Understanding human dimensions.* Washington D.C.: The National Academies Press.

NCDC (National Climatic Data Center). 2000. Billion dollar US weather disasters. http://www.ncdc.noaa.gov/ol/reports/billionz.html.

Okeil, A. M. and C. S. Cai. 2008. Survey of short- and medium-span bridge damaged induced by Hurricane Katrina. *Journal of Bridge Engineering* 13: 377–87.

Padgett, J. E., R. DesRoches, B. G. Nielson, M.Yashinsky, O.-S. Kwon, N. Burdette, and E. Tavera. 2008. Bridge damage and repairs from Hurricane Katrina. *ASCE Journal of Bridge Engineering* 13(1): 6–14.

Padgett, J. E., A. Spiller, C. Arnold, 2009. Statistical analysis of coastal bridge vulnerability using empirical evidence from Hurricane Katrina. *Structure and Infrastructure Engineering 2009: 1–11.*

Palmeri, C. Anger in Houston over post-Ike power outage. *Bloomberg Businessweek,* September 25, 2008.

Pasch, R., D. Brown, and E. Blake. 2005. *Tropical Cyclone Report: Hurricane Charley 9–14 August 2004.* Miami, Fla.: National Hurricane Center. http://www.nhcnoaa.gov/pdf/TCR-AL032004_Charley.pdf.

Paul, M. J. and J. L. Meyer. 2001. Streams in the urban landscape *Annual Review of Ecological Systems* 32:333–65.

Peacock, W. G., P. Maghelal, M.K. Lindell and C.S. Prater. 2007. *Draft: Hurricane Rita behavioral survey final report.* College Station, Tex.: Hazard Reduction and Recovery Center Texas A & M University.

Peacock, W. G., B. H. Morrow, and H. Gladwin, eds. 1997. *Hurricane Andrew: ethnicity, gender, and the sociology of disasters.* London: Routledge.

Perry, R. W. and M. K. Lindell. 1991. The effects of ethnicity on evacuation decision-making. *International Journal of Mass Emergencies and Disasters* 9(1): 47–68.

Perry, R. W. and A. H. Mushkatel. 1986. *Minority citizens in disasters.* Athens, Ga.: University of Georgia Press.

Pielke, R. A. and M. W. Downton. 2000. Precipitation and damaging floods: Trends in the United States, 1932–97. *Journal of Climate* 13: 3625–637.

Rappaport, E. 2005. *Addendum Hurricane Andrew 16–28 August, 1992.* National Hurricane Center. http://www.nhc.noaa.gov/1992andrew_add.html.

Rappaport, E. 2005. *Preliminary report Hurricane Andrew 16–28 August, 1992.*

National Hurricane Center. http://www .nhc.noaa.gov/1992andrew.html.

Resio, D. T. and J. J. Westerink. 2008. Modeling the physics of storm surges. *Physics Today* 61 (9): 33–38.

Schuster, W. D., J. Bonta, H. Thurston., F. Warnemuende, and D. R. Smith. 2005. Impacts of impervious surface on watershed hydrology: a review. *Urban Water Journal* 2: 263–75.

State of Florida. *2009 supplement to the 2007 Florida Building Code.* Tallahassee, Fla. http://www. ecodes.biz/ecodes_support/Free_ Resources%5CFlorida_Codes%5C2007_ Florida_Errata%5CPDFs/2007_FL_ BLDG_Chpt_1%20to%2012.pdf

State of Texas. 2009. Hurricane Ike Recovery Program: Assessment Data Sheet. Austin, Tex. Texas Department of Rural Affairs.

Stewart, S. 2004. *Tropical Cyclone Report Hurricane Ivan 2–24 September 2004.* Miami, Fla.: National Hurricane Center. http://www.nhc.noaa.gov/pdf/TCR-AL092004_Ivan.pdf.

Texas Cable News. 2008. Harris County: Mandatory evacuation for some— shelter in place for others. September 12, 6:46 a.m., CDT. http://www.txcn. com/sharedcontent/dws/txcn/houston/ stories/khou080911_jj_harris_county_ evacuations_Ike.66299b4a.html.

Tourbier, J. T. and R. Westmacott. 1981. *Water resources protection technology: A handbook of measures to protect water resources in land development.* Washington, D.C.: The Urban Land Institute:

TranStar. 2009. Technical working document, online hurricane surveys. Houston, Tex.

USDA (US Department of Agriculture). 2000. *Summary report: 1997 national resources inventory (revised December 2000).* Ames, Iowa: Natural Resources Conservation Service and Statistical Laboratory, Iowa State University.

US Department of Transportation. 2006. *Report to Congress on catastrophic hurricane evacuation plan evaluation.* Washington, D.C. http://www.fhwa.dot.gov/ reports/hurricanevacuation/rtc_chep_ eval.pdf

Watt, J. G. 1983. *Economic and environmental principles and guidelines for water and related land resources implementation studies.* Washington, D.C.: US Water Resources Council.

Whipple, W. 1998. *Water resources: A new era for coordination.* Reston, Va: ASCE Press.

Wilson, S. G. and T. R. Fischetti. 2010. *Coastline population trends in the United States: 1960–2008.* Washington, D.C.: US Census Bureau.

Zahran, S., S. D. Brody, W. G. Peacock, A. Vedlitz, and H. Grover .2008. Social vulnerability and the natural and built environment: a model of flood casualties in Texas, 1997–2001. *Disasters* 32: 537–60.

Yang, Z. and W. G. Peacock. 2010. Planning for housing recovery? Lessons learned from Hurricane Andrew. *Journal of the American Planning Association* 76(1): 5–24.

# Contributors

Philip Bedient, PhD, Professor of Civil and Environmental Engineering, Rice University

Jim Blackburn, JD, Environmental Lawyer and Professor in the Practice, Rice University

Tom Colbert, AIA, Associate Professor, University of Houston

Samuel David Brody, PhD, George P. Mitchell '40 Chair in Sustainable Coasts, Department of Marine Sciences, Galveston & Department of Landscape Architecture and Urban Planning, College Station, Texas A&M University

Clint Dawson, PhD, PE, Professor, The University of Texas at Austin

Nick Fang, PhD, Research Scientist, Rice University

Dustin Henry, GIS Analyst, City of Galveston, Texas

Carol Abel Lewis, PhD, Associate Professor and Director, Center for Transportation Training and Research, Texas Southern University

Jeffrey Lindner, Meteorologist, Harris County Flood Control District, Houston, Texas

Walter Gillis Peacock, Professor and Director, Hazard Reduction and Recovery Center Department of Landscape Architecture and Urban Planning, Texas A&M University

Himanshu Grover, AICP, Assistant Professor of Urban and Regional Planning, SUNY-Buffalo

Wesley Highfield, Research Scientist, Department of Marine Science, Texas A&M University-Galveston

Jamie E. Padgett, PhD, Assistant Professor of Civil and Environmental Engineering, Rice University

Jennifer Proft, PhD, Research Associate, The University of Texas at Austin

Hanadi S. Rifai, PhD, PE, Professor and Director of the Environmental Graduate Program, University of Houston

David C. Schwertz, Senior Service
Hydrologist, NWS Houston/Galveston

Antonia Sebastian, Research Assistant,
Rice University

Kevin Shanley, FASLA, CEO, SWA Group

Matthew Stearns, Research Assistant, Rice
University

Bill Wheeler, Deputy EMC, Harris County
Office of Emergency Management

Shannon Van Zandt, AICP, Assistant
Professor of Landscape Architecture &
Urban Planning and Coordinator, Master
of Urban Planning Program, Texas A&M
University

# Index